THE WASHINGTON PAPERS
Volume IV

38: Strategic Power: Military Capabilities and Political Utility

Edward N. Luttwak

THE CENTER FOR STRATEGIC AND INTERNATIONAL STUDIES
Georgetown University, Washington, D.C.

SAGE PUBLICATIONS
Beverly Hills / London

Printed in the United States of America

For information address:

SAGE PUBLICATIONS, INC.
275 South Beverly Drive
Beverly Hills, California 90212

SAGE PUBLICATIONS LTD
St George's House / 44 Hatton Garden
London EC1N 8ER

International Standard Book Number 0-8039-0659-5

Library of Congress Catalog Card No. 76-40529

SECOND PRINTING

When citing a Washington Paper, please use the proper form. Remember to cite the series title and include the paper number. One of the two following formats can be adapted (depending on the style manual used):

(1) HASSNER, P. (1973) "Europe in the Age of Negotiation." The Washington Papers, I, 8. Beverly Hills and London: Sage Pubns.

OR

(2) Hassner, Pierre. 1973. *Europe in the Age of Negotiation.* The Washington Papers, vol. 1, no. 8. Beverly Hills and London: Sage Publications.

CONTENTS

I. INTRODUCTION: STRATEGIC WEAPONS AND POLITICAL POWER

The great strategic debate continues. New weapons appear every few years and major new technologies emerge from the laboratories every decade or so, but the fundamental questions that were asked on the morrow of Hiroshima are with us still: how far can we safely rely on the delicate psychological mechanisms of deterrence? To what extent do we still need the costly assurance of non-nuclear forces? How much strategic power do we need to deter, say, an invasion of Europe—and how much less, if we seek only to avert a nuclear attack upon ourselves? And then there is also the question that many prefer not to ask: is it prudent to rely for our very survival on the offensive threat of deterrence alone, or can we build strategic defenses and provide some degree of protection for the civil population?

Each year we answer these questions (or merely shelve them) in the processes of Congressional decision. From the arguments of critics and supporters of the official policy, through the constant bureaucratic pressure of the services, in the annual review of the alternatives within the Executive, and from the shifting electoral calculations of senators and representatives the answers emerge piece by piece, often by inadvertence and rarely conceived in the whole. The decisions are quite properly dominated by elected officials and not by technical experts, for the fundamental questions are indeed not technical but political and even social. Quite properly also, all decisions are only provisional, apt to be revised each year in the dollars appropriated or withheld; decisions must remain provisional because the threat to be countered changes also, the technology evolves, and the implicit priorities of the nation are constantly redefined.

One thing alone is a fixed certainty: the process will continue. There are no ultimate weapons, there are no finite strategies, and there is no force that will permanently suffice. The competition for strategic power between the United States and the Soviet Union, between Russians and Chinese, and even between the major nuclear powers and the lesser, may accelerate or temporarily slow down—but it will not cease. Since the strategic competition is only the symptom of a much deeper and broader political struggle, it cannot be arrested by arms control treaties nor can it even be seriously constrained. The only effect of limitations on any one class of weapons, or any one mode of performance, is to stimulate new efforts to develop strategic power in directions as yet unconstrained.

Nor is there any hope of eventually circumscribing all directions of endeavor since the underlying base of scientific knowledge does not respect legalistic inhibitions in its unpredictable evolution. It may still be worthwhile to seek agreed restraints, or even to accept them unilaterally, because some avenues of competition are more damaging than others—as in the cases of the space, antarctic and atmospheric testing of nuclear devices successfully prohibited by treaty. But nothing is gained by the pretense that the malady can be cured through the suppression of its symptoms. Indeed, if it were possible to eliminate *all* nuclear weapons, truly forcing the genie back into the bottle, it seems very doubtful that we would choose to do it. Out of the terror of nuclear war there has come a peace so far unbroken between all powers possessed of nuclear weapons. If the nuclear competition were to be abolished, a now dominant competition for non-nuclear military power would take its place, at a higher cost (for those who must pay for their manpower) and at a greater risk, for then the stability derived from the universal terror of nuclear war would be prejudiced.

In the awesome presence of the nuclear sanction the most definitive, traditional forms of Great Power competition are foreclosed: the powers cannot test each other's strength in peripheral wars to reassess the balance of power, nor can they seek to reduce one another by devastating central warfare. But the struggle for control over the destinies of mankind has not ceased, and the nuclear forms of military power are not apolitical. Ideological ambition, the quest for security and the struggle for power over the lives of the nations, are all still pursued through the armed persuasion of the nuclear arsenals. The mechanism of nuclear suasion is manifest in the normal course of international politics: in Delhi as in Bonn, in Toyko as in Tel Aviv, political practitioners view the strategic power of others according to the medium of their own perceptions, and construe its meaning according to their wits.

In the permanent necessity of conciliating the demands of the strong, and in the hope of exploiting the weakness of the less strong, national leaders must seek to estimate the power of others, and their estimates, right or wrong, technically exact or utterly crude, will influence their conduct. It is thus through the uncertain medium of others' perceptions that strategic arsenals generate political power. To the extent that beneficial effects are evoked in the conduct of others, our strategic weapons produce a political return; to the extent that our own forces do not suffice to negate the images of power generated by the Soviet strategic arsenal, we pay a political price for failing to augment our strength.

Ordinarily we discount such third-party effects in our view of the functions of strategic weapons. Instead of a global perspective, we adopt one strictly bilateral in which only the two superpowers exist. Instead of reckoning with the full spectrum of armed suasion—latent and tacit as well as active and declared, supportive of our allies as well as coercive towards antagonists—we have customarily extended our recognition only to a narrow segment of the total spectrum: negative coercion, that is, deterrence. Moreover, we ordinarily assess this one form of coercive suasion only as against the Soviet Union, and more rarely against China. It is for this reason that we tend to refer back to the great crises of Cuba and Berlin whenever we consider the utility of nuclear forces. Accordingly, the deployment of nuclear weapons is not recognized as a form of power-accummulation in the world arena, but is rather seen as a form of insurance, to be activated only in a deep crisis. And since in this form of insurance the compensation paid to the policy-holders is reciprocal destruction, it is not surprising that many tend to dismiss the usefulness of nuclear forces, and of any increments to them.

Having first excluded from consideration all participants in international society but one or two at most, having then chosen to ignore all aspects of armed suasion but one, we naturally arrive at the conclusion that the optimal level of strategic power is the absolute minimum required for the direct deterrence of the Soviet Union (and possibly China). But of course such exclusions of reality are entirely unwarranted. Whether we extend our intellectual recognition to nuclear suasion in the whole or not, it exists. In the well-known episodes of deep crisis over Cuba or Berlin, for a few days or weeks the mechanisms of nuclear suasion were overtly manifest. But they did not emerge fully formed out of nothing. In a *latent* form the suasion was there all the time, continuously (if tacitly) shaping the conduct of all affected parties.

When national leaders are compelled to reconsider the strength of the strategic forces at their disposal in a crisis severe enough to present an actual risk of nuclear conflict, the question uppermost in their mind must be the extent of the damage which their societies could suffer if nuclear crisis becomes nuclear war. Without necessarily engaging in elaborate calculations, estimates are thus formed in men's minds of the outcome of reciprocal attacks, and these estimates will influence the degree of risk that leaders trapped in a desperate situation will be inclined to accept. Even when the possibility of nuclear conflict remains remote, as it always will unless the most vital interests are at stake, perceptions of the ultimate capabilities will limit the demands that national leaders will feel free to advance. Without seeking to predict the unpredictable interplay of move and countermove that might led to war, leaders aware of the ultimate risk and informed of the fundamental balance of strength must set tacit limits on their own conduct. Thus the remote contingency of war plays an important role in the ordinary course of international politics: in the presence of nuclear weapons strategic decisions may no longer be made without explicit recognition of the likely responses these decisions could evoke.

The limits on adversary actions set by *direct* nuclear deterrence are obvious enough: it is quite clear, for example, that the Soviet leadership could not order an all-out attack on American cities without facing the unacceptable prospect that many Soviet cities would be destroyed in turn. If this were indeed the *only* practical implication of the strategic balance, as some appear to believe, the formulation of strategic policy would indeed be a simple matter: it would suffice to deploy a small (but diversified) force capable of destroying, say, the 50 largest cities of the Soviet Union even after absorbing the full force of a "counterforce" attack against our deterrent. Once such a force is deployed, consisting perhaps of 10 quiet missile submarines, 100 of our present intercontinental ballistic missiles (ICBMs), and as many long-range bombers, we would merely have to make such adjustments to this force as the evolution of the Soviet counterforce potential might require.

A "minimum deterrence" force of 160 submarine-launched ballistic missiles (SLBMs) with as many warheads, 300 ICBMs with multiple warheads which can be targeted independently (MIRV), and the 100 bombers with their stand-off missiles, would be amply sufficient to guarantee the destruction of the 50 largest cities of the Soviet Union, even if a good many bombers, most ICBMs, and half the submarines should have been destroyed when the retaliatory attack is launched. Given the sharp increase in Soviet counterforce capabilities now taking place, an economical and survivable minimum deterrent force could still be deployed by placing the bombers on continuous airborne alert, by taking the ICBMs from their silos and placing them on mobile launchers, and relying more on the submarines. Why then do we not reduce our forces to the levels of minimum deterrence and leave them at that? If one fourth of our present SLBM force, one tenth of our ICBM force, and one third of our present operational bomber force is enough, why deploy more? And this indeed is the unspoken argument behind much of the by now institutionalized opposition to the strategic policy of every administration.

The plausibility of the minimum deterrence argument is deceptive. First, it totally ignores the political effects of nuclear capabilities. If we were to reduce our forces to the much lower levels required for simple (strike-back) deterrence, we would soon discover how tangible and important were the political benefits of larger nuclear forces—hard though it may be to define these benefits exactly in present circumstances; the political realm simply does not lend itself to exact difinition in the abstract. But if we were to engender the reality, we should soon discover that our allies—now exposed to the power of others and wholly unprotected by our nuclear forces—would soon part company with us to make the best settlement they could with those who have more than a minimum deterrent. And we would sooner still discover in the new freedom of action of our antagonists how important were those tacit restraints that our strategic power once imposed upon them. But this is after all implicit in the very argument: the strategy of a minimum deterrence is only intended to avert *direct* nuclear attacks upon the *cities* of the United States, and no other claims are made on its behalf.

If is obvious that no rational purpose could be served by a surprise attack against American cities by the Soviet Union. The act of destruction can gain nothing for an attacker, while it would trigger the one-shot response of the minimum deterrent. Thus the strategy provides protection against a threat which is exceedingly unlikely to materialize. It should be noted, however, that it also provides a degree of nuclear suasion, albeit negative: the retaliatory threat of minimum deterrence would suffice to guarantee that the Soviet Union could not extract concessions by threatening to attack our cities, for such threats would not be credible.

It is apparent, however, that once its demands on former neutrals and former allies are satisfied, the Soviet Union would nevertheless also seek to extract concessions from the United States. The minimum deterrent would of course foreclose attempts at nuclear blackmail based on threats against the American population. But the Soviet Union would have a perfectly acceptable alternative in the shape of *counterforce* threats. With a force only large enough to deliver a single strike against Russian cities, how would we respond to the gradual destruction at sea of our small 10-boat submarine force? We could scarcely retaliate by attacking Soviet

ICBMs or bomber airfields without inviting a still stronger counterforce attack against our own—an attack against which there could be no response short of attacks on Russian cities. Thus through the threatened use of its strategic weapons against our small minimum deterrent, bolstered perhaps by a few demonstration attacks, the Soviet Union could extract all the concessions which it could not gain by threatening our cities.

The inadequacy of the minimum deterrent, and its failure to provide any degree of armed suasion for the protection of the retaliatory force itself, reduce its value to very little. This is not merely a consequence of its small size, but also of the inflexibility inherent in a force only large enough to execute a single, catastrophic mission. That the minimum deterrence argument should still remain so persistent (witness the wide currency of the "overkill" metaphor to describe *all* larger forces) is not only a tribute to the political innocence of its advocates, but it also reflects the more widespread tendency to envisage nuclear strategy in the guise of warfare (rather than as an extension of international politics). And moreover this is a very special warfare in which there are only two moves: *their* all-out surprise attack, and *our* retaliation. It should be obvious that no all-out surprise attack on American cities by the Soviet Union can be a rational act, for by definition it could not be preceded by demands. The all-out surprise attack is the least realistic of threats, and should occupy a correspondingly reduced place in our strategic thought, except insofar as it is useful *to plan* on the basis of such attacks. (If strike-back forces can survive in that extreme circumstance, they will be adequate in any conceivable circumstances.)

Americans and Russians in positions of power are by now fully informed of the destructive capacity of strategic-nuclear weapons. Their use is likely to be very hesitant even in the most severe conditions, unless the command authorities themselves were to be destroyed. For this reason, strategic nuclear forces require the quality of flexibility needed by all military forces. Strategic weapons should be able to attack targets selectively and accurately; and the methods of command and control must be a good deal more flexible than the mythical red button.

For the United States, the requirement of operational flexibility for the strategic forces has a direct political significance. In order to protect the security of the United States in a wider sense that goes beyond physical survival, the limits which must be placed on Russian action must be much more restrictive than those imposed by direct deterrence. For example, the strategic nuclear forces must ensure by deterrence the protection of the so-called "forward-based systems" (FBS) deployed in Europe and in the Mediterranean. These are land-based and carrier-borne strike aircraft deployed for theater warfare but capable of delivering nuclear bombs into the Soviet Union. Their specific military task is to interdict base-areas and transit points in Eastern Europe and the western parts of the Soviet Union in the event of a Warsaw Pact invasion of NATO Europe which cannot otherwise be defeated. The FBS forces are of considerable political importance in U.S.-allied relations since they are a connecting link between American strategic-nuclear power and European defense by virtue of being the physical instruments of escalation. But the FBS are also rather vulnerable. The land-based aircraft are sheltered against conventional attacks but not, of course, against nuclear strikes upon their bases. So also for the F-4s, A-6s, and A-7s on board U.S. aircraft carriers within range of the Soviet Union. Its vulnerability, together with its peculiar importance, makes the FBS force the most likely victim of a limited counterforce strike in the event of an intense crisis. This is a threat that the FBS forces themselves cannot deter, and which must be averted by invoking the higher-level deterrence of the strategic-nuclear forces themselves.

The extended deterrence which serves to avert attacks upon the FBS also applies to a good many other assets which are also beyond the national territory, and which are also less than vital "core values" of the kind that a "minimum deterrent" may suffice to protect. But extended deterrence requires not only that the strategic forces be regarded as powerful in some generic sense but also that they be *flexible.* Over and above the base force needed for counter-city retaliation (which serves to deter attacks upon the population of the United States), an additional "disposable" strategic capability is therefore required. And while the base capability need not necessarily be very accurate or diversified in the yield of the weapons (though this is likely to make it more efficient), the "disposable" capability must have both attributes, not for the sake of efficiency but as an *absolute operational requirement.*

It would not, for example, be advisable to seek to deter a Soviet nuclear strike against the FBS by threatening retaliation against Soviet cities—which is all that a small force of low-accuracy strategic weapons could do. Such a retaliatory threat would lack credibility since it would be clearly disproportionate. And if then an attack upon the FBS *were* to take place, the choice of responses would be a choice of unacceptable evils: inaction—with terminal consequences for the Alliance, or else an attack on Soviet cities which would be bound to evoke a reply in kind. To deter an attack upon the FBS the retaliatory threat must therefore be aimed at *proportionate* Soviet values, such as Soviet tactical airbases, or storage depots for theater-level nuclear weapons. And if such deterrence were to fail nevertheless, the act of American retaliation though severe in the extreme would still be short of the utterly catastrophic.

The case of the FBS is one among many. In each instance values can be protected *if* there is the credible possibility of inflicting proportionate damage. In practical terms, this means that a strategic force with a wider range of weapon yields, with greater accuracies, and with a larger number of weapons, can offer (other things being equal) a greater coverage of extended deterrence, up to the point where *all* values on our side, and *all* worthwhile Soviet targets, are covered. On the other hand, if the weapons are inaccurate, so that it is difficult or impossible to launch a nuclear attack without inflicting widespread (disproportionate) damage; or if the strategic forces are too small to allow both a range of selective strikes *and* the retention of a counter-city deterrent in reserve, then extended deterrence will not be feasible.

It is for this reason that a much greater capability than that of a simple strike-back force is required in U.S. strategic-nuclear forces. And it is for this same reason that the upward adjustment in force levels, which is rendered necessary by each increment in Soviet *counterforce* capabilities, must be based not on the minimum deterrence baseline but on the much higher level of the strike-back force *plus* the disposable forces. Thus if progress in Soviet counterforce capabilities makes the *Minuteman* ICBM force vulnerable, we cannot simply rest content with the strike-back capability of bombers and missile submarines. For the detraction in U.S. strategic power would be tangible, entailing a corresponding loss of strategic options, and the concurrent creation of new Soviet options. Even if no attack on the Minuteman force were ever to take place, the change in the range of limited strike options open to the two sides would endow the Soviet Union with added political leverage. In a crisis, such leverage could be used overtly to extract concessions; more likely, in the normal course of U.S.-Soviet relations the awareness by each side of the *existence* of this particular Soviet capability would suffice to enhance the bargaining strength of the Soviet Union without any overt threats. In the presence of thermonuclear weapons, the diplomatic anticipation of possible actions in hypothetical circumstances can have the same impact on the course of international politics that the limited use of actual force might have had in the past.

It follows that the specifications of a strategic force for the United States are a good deal more exacting than those implied by the strategy of minimum deterrence. A simple counter-city retaliatory force is of course essential; while the particular and narrow threat it deters is exceedingly remote, this threat is also utterly catastrophic and ample insurance is fully warranted. But instead of setting a ceiling on the requirement, this level of capability is merely a minimal sine qua non. Over and above it, much greater additional capabilities are needed to provide as much "disposable" strategic power as the heavy commitments of the United States in Europe and elsewhere require.

It should be noted that there is no genuine parallel requirement for "disposable" capabilities on the part of the Soviet Union, as far as the Western Alliance is concerned. (China is another matter.) For the political circumstances of the two sides are utterly dissimilar: while the Soviet Union and its clients present a vivid and plausible threat to Western security and Western interest, there is no genuine reciprocal threat.

Official Soviet statements do make frequent reference to the warlike aims of the "imperialists," and from time to time Russian leaders still affect to believe that West Germany might present a revanchist threat to the Warsaw Pact, and even to the Soviet Union itself. But it is impossible to accept such statements as accurate reflections of Soviet beliefs. The nature and disposition of NATO military forces, the tone of political life in Western Europe, and the fundamental direction of American conduct are utterly inconsistent with warlike intentions towards the Soviet Union and its periphery of client-states. By contrast, the deployments of the Warsaw Pact present the classic profile of an offensive striking force aimed at West Germany and the English Channel; indeed the Soviet Union has derived great benefit from this offensive deployment which gives it a degree of influence-by-intimidation over Western Europe which neither its provincial culture, nor its unattractive society, nor its primitive economy could have generated.

Deficient as it is in all the normal attributes of Great Power status except sheer geographic extent and population, the Soviet Union is a superpower by virtue of its military strength alone. To the extent that Soviet military capability is negated by that of the United States and its allies, the Soviet Union would be correspondingly reduced to a level of influence which more closely approximates its true entitlement. At present, the combination of U.S. strategic strength and NATO theater-level capabilities is plainly insufficient to do this. If it were adequate, the military equation would be fully balanced and this would allow East-West relations to be governed by the interplay of *positive* national attributes, cultural, social, and economic.

Returning to the strategic-nuclear dimension of the problem, we find that it is impossible to define in the abstract the precise requirements of both elements in U.S. strategic capabilities. It is a fairly straightforward exercise to estimate the counter-city requirements of simple deterrence: one begins by estimating the number and type of warheads needed to destroy the requisite square miles of Soviet urban area, and then one adds whatever is needed to offset the estimated losses which Soviet counterforce capabilities could inflict, as well as to compensate for technical faillures and maintenance cycles. But the requirements of "disposable" capabilities are not amenable to such simple analysis. Instead it is necessary to investigate in full detail the range of distinct values which require protection (civil and military assets of allies plus non-urban U.S. assets, primarily military), and then to evaluate the range of "proportionate" Soviet targets. Between them, the two define the choices of limited strike options which are both necessary *and* credible in their deterrent intent. And then we should add the compensating increments needed to offset Soviet counterforce capabilities as well as maintenance and unreliability factors. A very rough comparison of some of the relevant data is presented in chapter III, but only a full scale planning exercise can be adequate to the task.

It is important to take note of a further desideratum for American strategic forces, which may or may not entail additional physical requirements in the deployment of American strategic forces. This desideraturm derives from the more general suasion which emanates from third-party perceptions of strategic power, as they affect countries other than the Soviet Union or the United States. It is impossible to determine how the political leaders of the lesser powers measure strategic strength. Certainly their perceptions are not based on detailed numerical studies. More likely, their views of the strategic balance are impressionistic and based on images which may be largely qualitative in nature. Thus long-range bombers may be regarded as "old-fashioned" merely because they are conventional manned aircraft and regardless of their effective capabilities in real combat environments. Equally, the deployment of heavy ICBMs such as the SS-9 or SS-18, which have been repeatedly described as "super-missiles" or "giant ICBMs," may make a deep impression independent of their actual military capabilities.

It is now known that the clear-cut differences in Soviet and American missile force-level ceilings under the provisions of the 1972 SAL-I "Interim Agreement" generated vivid images of Soviet strategic superiority (witness, for example, the statements in both the British and the West German defense "White Papers"), while it is also apparent that the countervailing American advantages in accuracy and MIRV subdivision went almost unnoticed. Even the most casual readers of the daily press learned by constant repetition that the United States was only allowed to deploy 1,054 ICBMs as against 1,618 for the Soviet Union, and that the maximum limits on SLBMs were 710 for the United States and 950 for the Soviet Union. It was the far smaller number of technically qualified observers who were aware of the American advantage in bomber capabilities, in missile accuracies, and in the number of separately deliverable warheads.

While the much-publicized SAL-1 force levels penetrated the consciousness of informed opinion the world over, the strategic significance of the offsetting factors had no such impact. Non-expert observers, a category which certainly includes all but a fraction of the world's political elites, accordingly came to the conclusion that the United States had chosen to concede a position of overall military superiority to the Soviet Union. It is important to realize that this assessment could be made *even* if the offsetting American advantages were given due recognition. This because the Soviet advantages in SAL-1 were fixed and contractual, while most U.S. advantages were contingent on a technical superiority that was bound to be eroded in time (and the U.S. advantage in MIRV subdivision and accuracies has indeed been eroded). Moreover, third-party observers could reasonably conclude that the Soviet Union was militarily superior *even* if the strategic strengths of the two sides were and remained roughly equal, because of the prior superiority of Soviet *non*-nuclear forces.

For many national elites, the import of these perceptions (compounded as they were by the equally vivid impression made by the domestic disarray of those years) was that the time had come to conciliate the Soviet Union, whose long-established preponderance in non-nuclear strength was about to be augmented by a degree of strategic superiority also.

It is in this context that the terms of the Vladivostok agreement-in-principle must be seen. Advocates of arms control have complained that the possible force-level ceilings (2,400, or perhaps 2,200 weapons on each side) are much too high. But there is no doubt that the formal numerical equality of the two strategic forces is of major political importance. The reversal of the formal inequality of the SAL-1 force-levels is an urgent necessity—even though with such widely different weapons deployed on the two sides it is not at all clear that de jure numerical equalities would translate into de facto equal capabilities.

It is interesting to note that the masters of American diplomacy have presented the equality in force levels agreed at Vladivostok as a major negotiating achievement. A mere two years earlier the same negotiators had insisted that the unequal ceilings of SAL-1 were of no significance whatsoever. Due recognition should also be given to the high skill of Soviet diplomacy: having received higher force-level ceilings in 1972 as a free concession, the Russians were able to exact a price for the numerical parity agreed at Vladivostok. It is possible, however, that this time the Russian negotiators have overplayed their hand, since the concessions they demanded—and especially the prohibitions on long-range cruise-missile deployments—are likely to be unacceptable to the U.S. Congress. Discussed in detail in chapter II, the tentative Vladivostok guidelines for a ten-year SAL treaty may therefore prove to be abortive.

In what follows, the strategic forces of the nuclear powers are described and some aspects of their capabilities are evaluated. But as the analysis becomes more technical, it is important to bear in mind that the critical dimension of strategic policy is not technological but political, and that the necessarily vague perceptual factors may in the end count for much more than the detailed estimates of weapon performance.

II. THE ANATOMY OF STRATEGIC-NUCLEAR POWER

The Soviet Union and the United States deploy superficially similar "triads" of nuclear delivery forces capable of reaching each other's homelands, but there are profound differences in their composition, and there is less than meets the eye in their apparent similarities. In addition to the land-based ICBMs, long-range bombers and submarine-borne ballistic missiles (SLBMs), the United States and the Soviet Union also have a good many more long-range nuclear weapons that are officially regarded as non-strategic (even though they could reach homeland targets) because they are qualitatively inferior or performance-limited in some way. These include Soviet medium bombers which could reach the United States on one-way missions; American tactical strike aircraft deployed on Navy carriers and in Europe capable of reaching the parts of the Soviet Union with nuclear weapons; Soviet ballistic missiles on board non-nuclear submarines; and Soviet long-range cruise missiles on board both nuclear and non-nuclear submarines. Additionally, the Soviet Union deploys a large but now rather antiquated force of intermediate range (IRBM) and medium-range (MBRM) ballistic missiles targeted on Western Europe or China.

Among the lesser nuclear powers, only France deploys a full "triad" of offensive forces, albeit in miniature: four ballistic-missile submarines, 18 IRBMs, and 36 light bombers all of which are targeted on the Soviet Union. Britain still retains some 50 "V" bombers which could attack the Soviet Union with nuclear weapons, but its only official strategic force consists of four ballistic-missile submarines. China has not yet deployed submarines in its strategic arsenal but has a number of medium bombers and a growing force of land-based missiles targeted on the Soviet Union. In addition, many Russian targets could be reached by Chinese fighter-bombers capable of delivering nuclear weapons. India has for the time being only a handful of plutonium bombs which could no doubt be delivered on Chinese and Pakistani targets by light bombers or transport aircraft. There is as yet no evidence of an Indian development effort aimed at producing more reliable delivery vehicles.

There is an obvious imbalance between these vast offensive forces and the rather weak strategic defenses deployed on all sides, an imbalance that is historically unprecedented. In particular, no substantial missile defenses are deployed anywhere, even though the ballistic-missile defense technology available to the United States—and perhaps the Soviet Union—is by now at least as effective as that used in anti-bomber air defenses—which *are* deployed. Under the terms of the 1972 AMB Treaty as amended in 1974, the Soviet Union and the United States may each deploy only a single and spatially limited missile-defense complex with no more than 100 nonreloadable interceptor missile launchers. No such treaty limitation applies to all the other countries that could become the victims of ballistic-missile attack, but in their case there are organic limitations, since the cost of ballistic-missile defenses is high and the technology is very complex.

Large air-defense forces consisting of ground-based radar and control networks, manned fighter interceptors, anti-aircraft missiles, and guns remain in service in the Soviet Union and China; smaller air-defense forces based primarily on manned interceptors are deployed by the United States, Britain, France, and many other countries. Another form of strategic defense, though not officially labelled as such, is anti-submarine warfare. Large tactical anti-submarine (ASW) forces are deployed by the major powers, and there is no fundamental distinction between "tactical" ASW for shipping-protection and "strategic" ASW aimed at destroying SLBM forces at sea. Operationally the employment of the forces would be quite different, but the assets useful for tactical ASW would be equally useful for strategic ASW. Britain, the Soviet Union, and the United States all deploy nuclear-powered attack submarines, the most effective single ASW weapon, but any ASW campaign would have to be a coordinated effort involving land-based patrol aircraft, surface warships, and sea-bed sensor grids as well as attack submarines. Thus the ASW forces maintained by all significant naval powers have an inherent strategic potential. However, so long as the principal means of location remain ineffectual against well-built and well-handled nuclear submarines, the role of ASW forces in the strategic equation will continue to be of small significance.

Land-Based Ballistic Missiles

Still perhaps the strategic weapon par excellence, the fixed-site ballistic missile is now slowly losing its primacy. On the one hand, submarine-launched ballistic missiles now offer a more enduring guarantee of survivability, since increasingly accurate warheads are gradually undermining confidence in the protection which steel and concrete underground launchers ("silos") can provide. At the same time, the new-style lightweight cruise missiles developed by the United States now offer the possibility of attaining extreme accuracies, far exceeding those of ballistic missiles guided by present inertial devices. In fact cruise missiles may eventually render the silos of ballistic missiles highly vulnerable even to *non*-nuclear attack.

In theory, it would be possible to replace the unguided warheads of present ICBMs with terminally-guided maneuvering warheads capable of great precision. But such devices would be much more complex, much more costly, and more vulnerable to countermeasures than equally accurate cruise missiles: ballistic missiles reach their targets in high parabolic trajectories which in full-range ICBMs reach hundreds of miles into space; cruise missiles, by contrast, fly towards their targets at very low altitudes, and use automatically-recognized terrain data to make repeated flight-path corrections. On the other hand, so long as the 1972 ABM Treaty between the United States and the Soviet Union remains in force, ballistic missiles do not have to contend with interception, while cruise missiles could in theory become highly vulnerable though only if very-low altitude air defenses are deployed, at enormous cost. (It should be noted that the ABM Treaty does *not* prohibit accuracy-degrading countermeasures against terminally-guided ballistic warheads.)

While the eventual demise of the ICBM is already in prospect, it still remains the most important weapon in the Soviet strategic arsenal, and perhaps in the American as well. In the official listing of strategic inventories ICBMs account for 1,500 out of the total of 2,705 Soviet delivery systems, and for a more than proportionate share of the 3,500 operationally-ready warheads and bombs in the Soviet strategic force. In the American arsenal, ICBMs account for 1,054 out of 2,319 delivery systems but their share of the warhead total is only 2,100 out of 8,500; on the other hand, ICBM warheads are both more powerful and more accurate than the more numerous SLBM warheads.

The numerical superiority of the Soviet ICBM force reflects the sharply divergent strategic policies which the two superpowers have followed over the last decade and a half. For all the popularity of the "arms race" metaphor, it is clear that there has been no two-sided race as far as the deployment of ICBMs is concerned:

TABLE 1
American and Soviet ICBM Inventories

	1962	1964	1966	1968	1970	1971	1972	1973	1974	1975	1976
U.S.S.R.	75	200	300	800	1,300	1,510	1,527	1,527	1,575	1,600	1,500
U.S.	294	834	904	1,054	1,054	1,054	1,054	1,054	1,054	1,054	1,054

[These rough estimates include ICBMs undergoing conversion.]

The unites States unilaterally stopped "racing" by 1964 at the latest, while the Soviet Union began to race in earnest *after* it had become evident that the United States would not deploy more than 1,054 ICBMs.

Soviet and American ICBMs include a variety of very different weapons and the two forces embody radically different conceptions of strategy. The present American force consists of 1,000 Minuteman-type missiles, small (56-60 feet high, 5.5 feet maximum diameter) solid-fuel ICBMs with correspondingly small payloads ("throw-weights") on the order of 2,000 pounds or so. The remaining 54 missiles are much larger liquid-fuel *Titan-2s,* originally deployed in 1962 and the only large ICBMs retained in the American force. The product of a managerially-efficient strategic policy which was almost untouched by foreign policy considerations, the American ICBM force was designed to launch *small* warheads, optimized for the counter-city role prescribed by a doctrine of simple deterrence. By 1964 at the latest, the decision had been made to halt the numerical growth of the force and to limit further progress to qualitative improvement. This was manifest in the single-warhead Minuteman-2 which was first deployed in 1965 and then much more fully in the Minuteman-3, the first ICBM to be equipped with a MIRV system. In other words, the American ballistic missile program was restricted to one particular type of weapon, the small economical solid-fuel ICBM.

A research program for large ICBMs (WS.120) was terminated, as were all programs for medium- and intermediate-range missiles (MRBMs/IRBMs). While the latter were inherently second-best weapons for the United States given its geographic position, the cancellation of the large-ICBM project reflected the planners' failure to appreciate the value of *flexibility* in the ICBM force, itself the consequence of a broader lack of understanding of the role of strategic weapons in foreign policy. Nothing illustrates the essentially apolitical conduct of U.S. strategic policy better than the precipitate withdrawal of older ICBMs: 63 Titan I, 143 Atlas and more than 700 Minuteman I ICBMs were withdrawn from service during the 1960s. This satisfied the USAF's striving for the modern, the system analysts' striving for efficiency, and the arms control hopes of the then-Secretary of Defense, but made no allowance for *diplomatic* considerations. In contrast, the Soviet Union still retains in service all the ICBMs it has ever built, with the exception of fewer than 50 very primitive and hopelessly unreliable SS-6 ICBMs, and perhaps 100 SS-7s. The latter have been withdrawn only very recently, and not *ex gratia*: under the provisions of the SAL-1 accord, the Soviet Union was allowed to add up to 209 SLBMs to its force in exchange for the withdrawal of older ICBMs.

American defense planners may have regarded Soviet deployment practices as primitive since gross numbers did not figure in *their* concepts of efficiency. But it is evident that "mere numbers" in fact determined the force levels allowed under the SAL-1 accord, resulting in the higher ceilings allowed to the Soviet Union. Had the Minuteman-1s been retained, the United States would have had as many ICBMs as the Soviet Union in 1972. By wisely according a higher priority to diplomatic visibility than to mere technical efficiency, Soviet strategic policy thus gained a significant diplomatic advantage.

The American ICBM program of the 1960s also ignored traditional military priorities. While managerial efficiency naturally leads to a search for the optimal system, which is then to be mass-produced to attain low unit costs, military needs and ordinary prudence call for diversification. Producing three or four different types of ICBMs necessarily results in higher unit costs but it also tends to produce a more versatile total force and one much more resilient to technical failure.

While the United States concentrated the full range of its technical advances into the Minuteman, the Soviet Union has developed three distinct classes of ICBMs: "light," "medium," and "heavy." Moreover, the Soviet force is technically as well as militarily diversified, since the "light" missiles are propelled by solid fuel, while the medium and heavy ICBMs have liquid-fuel engines. It is possible that Soviet conduct reflects a simple-minded projection of tradtional artillery practices into the ICBM era. But it should be noted that each class of ICBM in the Soviet force is distinctly suitable for a particular range of missions.

Thus the light solid-fuel ICBM class is inherently suited for mobile deployments on road or rail vehicles, or (packed in canisters) for air-launch from large transports (which can be kept on airborne alert to defeat ICBM attack). Originally represented by the SS-13 first deployed in 1969, this class now has a follow-on missile, the much-improved SS-16. Both of roughly Minuteman size, the SS-13 is a single-warhead ICBM much inferior to the Minuteman-3 but the SS-16 has a larger throw-weight and has been tested with a MIRV system. Both ICBMs would be especially suitable for attacks against cities.

All the medium ICBMs like the heavies are propelled by liquid-fuel engines. Originally represented by the SS-11 which was first deployed in 1966, the medium class is now present in two SS-11 variants and in two entirely different ICBM types, the SS-17 and SS-19. All these missiles are physically larger than the Minuteman type but while the first SS-11 variant is inferior in overall performance to the Minuteman-2, the SS-17 is much more accurate and has a substantially greater throw-weight (now utilized for a four-warhead MIRV system); the SS-19 is larger still and it too has a MIRV postboost vehicle (six warheads); it appears to be more accurate than the SS-17. On the other hand, the latter is being deployed in "cold-launch" silos, in which the ICBM is ejected by gas pressure before booster ignition. As well as facilitating the construction of better-protected silos, this technique allows bigger ICBMs to be accommodated and makes it possible to reload silos without the extensive refurbishing needed after a "hot" launch. Both the SS-17 and SS-19 must be considered as potentially much more capable than the Minuteman-3, even though the latter may still be more accurate at present. (Post deployment accuracy upgrading is to be expected in both the SS-17 and the SS-19.)

It is characteristic of Soviet practices that two separate ICBM types have been deployed in this one class of weapons; equally revealing is the fact that while SS-17s and SS-19s began to replace the older SS-11s in the course of 1975, two newer models of the latter were still being deployed in newly-built silos (identified as being under construction in the SAL-1 accords). Evidently the Soviet program is proceeding on its own industrial timetable, and it is deemed prudent to deploy these interim weapons rather than to convert the silos for SS-17s or SS-19s.

The heavy class of Soviet ICBM has presented serious problems to American defense planners and well-intentioned arms' controllers alike. While the "medium" ICBMs are evidently the general-purpose weapons in the force, and only acquired a "counterforce" profile when the SS-17 and SS-19 appeared in 1974-1975, the heavy SS-9 could never be seriously considered as an ordinary ICBM meant primarily for counter-city use.

By the time the first SS-9s were deployed in 1966 it had become abundantly clear that the United States would not develop a large ICBM, on the grounds that large warheads would threaten Soviet ICBM silos and thus undermine the stability of "mutual" deterrence. This decision was made *before* the advent of MIRV, which had ambiguous consequences from the arms control point of view. On the one hand, MIRV systems could be used to subdivide the throw-weight of small Minuteman-type missiles to yield still smaller warheads. This would of course serve to reinforce "the stability of deterrence": with the accuracies then available such small warheads could not threaten Soviet ICBM silos. At the same time, the multiplicity of warheads would serve to defeat population-defense ABM systems, thus removing another threat to stability (since an effective ABM would prejudice retaliatory capabilities, and therefore deterrence).

But it was always realized that MIRV systems could have opposite arms' control consequences if fitted on large (and accurate) ICBMs: instead of subdividing their throw-weight into small warheads suitable only for counter-city use, they could *multiply* counterforce effectiveness by enabling each large ICBM to destroy several silos. Everything thus depended on the type of ICBM on which MIRV systems were deployed: a MIRV system fitted on a very large ICBM of the SS-9 type would clearly fall in the second category. Only the hypothesis of a wholly uncontrolled industrial program conducted without reference to higher authority can explain away the counterforce intent of the Soviet heavy ICBM development program.

First deployed in 1966 the SS-9 has since appeared in several variant models; over 250 missiles of this type remain in service, fitted with single warheads whose yield is *more than 100 times as great as that of the Minuteman-3 warheads*. Hopelessly wasteful if used against cities (that is, for simple deterrence) these warheads are suitable for attacks on command centers and ICBM solios. Now a follow-on heavy ICBM has appeared in the SS-18; no larger than the SS-9 this missile nevertheless has a greater throw-weight, and one of its variants is fitted with a MIRV vehicle for eight warheads, each suitable for counterforce use. Another has 12 warheads, also large.

For a brief period in 1962 Secretary of Defense Robert McNamara publicly considered the possibility of targeting military installations rather than cities in a first retaliatory response. By 1964 the notion had been definitely abandoned. It is evident, however, that Soviet planners have never accepted the validity of the arms' control theories that dominated U.S. strategic policy during the 1960s, preferring instead to pay the high costs of acquiring a counterforce capability.

When the SAL-1 agreements were being negotiated, the American side attributed great importance to establishment of a sublimit on heavy ICBM launchers. In the light of what has happened since then, it appears that the value of this limitation was much exaggerated. Even though the heavy SS-9s and SS-18s are constrained to a maximal level of 308 units, the SS-19 deployed since then is *not* included in the sublimit, even though it is a much larger weapon than the SS-11 (100 viz. 70 cubic meters), and has a throw-weight three and a half times as great as that of the Minuteman. If, as it has been suggested, the United States had to make significant concessions to obtain the sublimit on heavy ICBMs—on the assumption that only the latter had a serious counterforce potential—it is apparent that the outcome has been disappointing.

The larger size of Soviet ICBMs together with their numerical superiority has naturally resulted in a gross imbalance in the overall throw-weights of the two ICBM forces. At present, overall Soviet ICBM throw-weight amounts to roughly five million pounds versus approximately two million pounds for the United States. As the follow-on ICBMs replace older weapons in the Soviet force, the gap is expected to grow over the next decade. A middle-range throw-weight estimate for the 1985 Soviet ICBM force would be approximately 10 million pounds (308 SS-18s at 15,000 pounds; 1,032 SS-17s and SS-19s at an average of 5,000 pounds); by contrast the U.S. total will have declined to under two million pounds.

The throw-weight of a missile force merely delimits its ultimate potential; it does not define its actual capabilities. In practice, to convert their potential advantage in missile throw-weight into an actual superiority in strike capabilities, the Russians would need only a combination of accuracy and weapon yield-to-weight ratios which is more than one fifth as effective as that of the American force. Further, in order to distribute this overall strike power efficiently as between thousands of targets, multiple warheads systems are required. It is very hard to estimate how much energy (destructive power) Soviet warheads actually yield. (The margin of uncertainty for a typical late-model warhead is 0.5-1.5 megatons.) It is still more difficult to estimate missile accuracies, in fact not even the Russians can be quite certain. But because of their throw-weight advantage, Soviet planners face a relatively undemanding task as compared to the American, and there is no reason to believe that Soviet military technology is not making rapid progress in this area. (The wide technology gap between the civilian economy of the Soviet Union and that of the United States is irrelevant: the self-contained and privileged military industries of the Soviet Union are known to be provided with technology far more advanced than that available to the civilian industry.)

It is interesting to note that in the one area of missile capability where our data is unequivocal, Soviet progress towards exploiting the potential advantage in throw-weight is a demonstrated fact: operational MIRV post-boost vehicles with on-board computers are present in no fewer than four different types of ICBMs—the SS-16, the SS-17, SS-19, and SS-18—all operational by 1975.

Table 2 presents the author's estimates of Soviet and American missiles characteristics. It will be noticed that the single most important variable in determining the strike power of the two forces is missing: accuracy. Some accuracy estimates are given in chapter III; these are inherently much more uncertain than other performance estimates.

Such technical uncertainty would count for far less if we knew more about the basic objectives, planning constraints, and cost/benefit calculations of Soviet strategic planners. While the United States publishes

TABLE 2

Salient Characteristics of American and Soviet Land-Based Ballistic Missiles

		Year IOC*	Status**	Throw-weight 'ooo pounds	Warhead Systems Type	Number	Yield***	Remarks
United States								
Minuteman-2		1965	OX	2.5	Single	1	1 Mt.	450 are in service. Some may be replaced.
Minuteman-3		1970	O	2.5	MIRV	3	170 Kt.	New 300-350 Kt. warheads in production.
Titan-2		1963	OW	8.0	Single	1	9 Mt.	To be replaced by Trident SLBMs.
Soviet Union								
SS-7		1962	OW	7+	Single	1	5 Mt.	Being withdrawn in exchange for SLBMs.[b]
SS-8		1963	OW	7+	Single	1	5 Mt.	Only 19 originally built. As above.
SS-11	Mod. 1	1966	OX	2.0	Single	1	1-2 Mt.	Only Soviet ICBM to be mass-produced.
	Mod. 2	1973	OL	2.0	Single	1	1-2 Mt.	Has penetration aids. In new silos.[c]
	Mod. 3	1973	OL	2.0	MRV[a]	3	3-400 Kt.	(Not MIRV) A few have been deployed.
SS-13		1969	OX	1.0	Single	1	1 Mt.	First solid-fuel Soviet ICBM.
SS-16		1976	E	2.0	MIRV	—	—	May or may not replace the 60 SS-13s.[d]
SS-19		1974	O	6.0	MIRV	6	0.4-1.5 Mt.	Under one-megaton warheads likely.
SS-17		1975	O	4.0	MIRV	4	0.5-1.5 Mt.	Under one-megaton warheads likely.
SS-9	Mod. 1	1967	OL	10-15	Single	1	15-20 Mt.	Few have been deployed.
	Mod. 2	1966	OX	10-15	Single	1	20 Mt.+	Standard heavy ICBM.
	Mod. 3	1969	E	10-15	Single	1	?	Used for FOBS experiments.
	Mod. 4	1971	E	10-15	MRV	3	5 Mt.	(Not MIRV) None are deployed.
SS-18	Mod. 1	1974	OL	15	Single	1	20 Mt.+	Being deployed, replacing SS-9s.
	Mod. 2	1975	O	15	MIRV	8	1 Mt.+	Other warhead combinations also.
	Mod. 3	1975	E	15	Single	1	15 Mt.?	Accuracy-enhancement testing.
Medium and Intermediate-range Ballistic Missiles								
SS-4		1959	1000-mile range MRBM		with megaton-range			Single Warhead
SS-5		1961	2000-mile range IRBM		with megaton-range			Single Warhead
SS-20		1976	3000-mile range IRBM		with MIRV[e]			Silo and mobile-deployed

NOTES ON TABLE 2:

*Year IOC = initial operational capability (i.e., deployed silos in operational launchers).

**Status = O = operational in significant numbers.
OX = operational in significant numbers but may be withdrawn.
OW = withdrawal planned for near term.
OL = operational but in small numbers.
E = experimental weapon; not operationally deployed.

***Yield: in million metric tons of TNT equivalent = megaton (= Mt.); or in thousand metric tons of TNT equivalent = kiloton (= Kt.).

(a) MRV = multiple ("cluster") warhead system, where the individual re-entry vehicles fall in a fixed pattern and cannot be separately aimed.
(b) Some SS-7s and SS-8s are still deployed in soft launchers. Both are being exchanged for new SLBMs pro rata; the ICBMs are supposed to be dismantled when additional SLBM tubes appear on submarines in sea trials. Overlaps have been experienced.
(c) Mod. 2 and Mod. 3 SS-11s have been deployed in silos which were listed as being under construction in 1972.
(d) The SS-16s may replace the SS-13s (60 in all); or they may be deployed in land-mobile configurations; or they may remain experimental. The SS-20 IRBM consists of two stages of the three-stage SS-16.
(e) With a 1.5 Mt. warhead the range exceeds 3,000 N. miles. With a smaller Kiloton-range warhead it attains 4,000 N. miles. A SALT violation may be entailed since some SS-20s have been deployed in new silos built within the range of the United States.

---

precise details of its strategic programs years in advance, giving extensive background information on the driving considerations behind each decision, the Soviet Union publishes nothing. Nor have the long years of supposedly intimate SAL negotiations shed much light on Soviet strategic policy. In spite of many attempts by the American side to initiate a general exchange of ideas on strategic matters, Soviet negotiators have consistently refused to go beyond the arms' length exchange of specific proposals. As a result, we still know nothing about Soviet stragegic policy except what can be learned by satellite photography and electronic intellitence. Enormously useful though this is, it does not tell us enough about the present, and much too little about the future.

Owing to the more advanced state of American guidance and warhead technologies, which result in a significantly more efficient use of the throw-weight than in present Soviet ICBMs, it is sometimes suggested that the Minuteman-3 is also potentially a counterforce weapon, and thus in some sense comparable to the SS-18. It is of course true that with *very* high accuracies even small warheads such as those now fitted on Minuteman-3s could destroy Soviet ICBM silos. In fact, with the extreme accuracies now commonly achieved in certain short-range tactical missiles it would be possible to destroy heavily protected ICBM silos even with warheads much smaller than those of the Minuteman-3, or indeed with conventional explosives. So far, no real attempt has been made in the United States to achieve such extreme accuracies. The accuracy-enhancement development efforts now being pursued have the much more modest goal of achieving software refinements in existing guidance systems, rather than producing entirely new ones.

But comparisons between the Minuteman-3 and the altogether larger SS-18 are also misleading on more fundamental grounds. In calculating missile kill-probabilities against small and protected targets such as missile silos, the conventional counterforce formulae combine accuracy and warhead yield data. But accuracy and yield are in fact *qualitatively* different. While a warhead of a given size will invariably produce a certain energy yield when it explodes, unless there is a total breakdown, accuracy is inherently subject to degradation from many causes, some very hard to predict at the moment of launch. The conventional measure of missile accuracy, the CEP (circular error probable) measures only the expected median *in*accuracy, on the basis of past tests. Each side *knows* the yield of its own missile warheads but can only *estimate* their accuracy. To obtain good accuracy estimates large numbers of ICBMs picked at random would have to be lauched from operational silos and in the direction of the real target. For financial reasons, present CEP estimates for American ICBMs are instead based on data obtained from a small number of test firings, mainly from special silos and none aimed at the Soviet Union.

Hence, the more the effectiveness of ICBMs against hard targets derives from yield rather than accuracy, the greater is the confidence that may be placed in their eventual performance. It is revealing that the one ICBM development program now taking place in the United States which *is* explicitly directed at increasing counterforce effectiveness provides for an increase of warhead yields. New Mk. 12A warheads with an estimated yield of roughly 350 kilotons are to replace the present 170 kiloton warheads in some of the Minuteman-3 MIRV systems. It should be noted that even this increased yield only amounts to one third of the probable yield of each of the eight warheads of the SS-18 MIRV version.

Under the provisions of the SAL-1 accords, the number of ICBM silos (not the ICBMs themselves) was limited to those operational or deemed to be under construction in May 1972. The Soviet Union refused to include actual numbers in the text of the Interim Agreement. Instead, it was understood that U.S. estimates would be accepted (the use of satellite reconnaissance was legitimized in the accords). Three categories of launchers were defined in the Agreement:

(A) Silos for "old" (pre-1964) ICBMs. These silos could not be modernized. If not retained in service, they could be exchanged on a one-for-one basis for SLBMs, with the Soviet Union entitled to 209 new SLBM units (vice SS-7s and SS-8s) and the United States to 54 units (vice Titan-2s).

(B) Silos for "light" ICBMs. The allowed units were 1,000 for the United States and 1,030 for the Soviet Union, and were actually in use for Minuteman-2s and -3s, SS-11s, and SS-13s respectively. In addition, 61 light silos were accepted as being under construction in the Soviet Union, resulting in allowable totals of 1,000 and 1,091 respectively for the United States and the Soviet Union.

(C) Silos for heavy, modern ICBMs. No exact—or even meaningful—definition of a "heavy" ICBM was stated, hence the SS-19 problem. The United States was allowed no silos in this category; the Soviet Union was deemed to have 288 operational, with another 20 under construction (the former being used for the SS-9, the latter now for the SS-18 Mod. 1).

The actual inventories comprised under these categories are estimated in the first column of Table 3, with justifications appended. Since the SAL-1 accords were negotiated in 1972, no new agreements limiting offensive weapons have been signed. However, the outlines of an agreement-in-principle were published in the wake of the Ford-Breshnev meeting held in Vladivostok in November 1974; if the SAL-1 Interim Agreement is replaced by a more lasting treaty by the time it expires in October 1977, it is reasonable to expect that the new accord will incorporate the Vladivostok principles.

These provide for an overall weapon ceiling of 2,400 or possibly 2,200 units. While full definitions of what is and what is not a "strategic weapon" have *not* been agreed, it is clear that all ICBMs would be included. There are in addition two subordinate constraints: the sublimit on heavy, modern ICBMs carried over from SAL-1, with 308 allowed to the Soviet Union and zero to the United States (not that there is any reason for the United States to invest in heavy ICBMs) and, a sublimit of 1,320 units on MIRVed strategic missiles, and perhaps also on bombers armed with long-range cruise missiles. In Table 3 projections of the two ICBM forces are given under these prospective SAL rules. For this purpose it seemed convenient to assume the overall weapon ceiling at 2,400 units rather than at the somewhat lower figure which is equally probable. Further while the United States would, of course, use its MIRVed weapon allowance for both, SLBMs and ICBMs, it is assumed that the Soviet Union would use all its units for ICBMs.

On the American side the calculation is straightforward. In 1985, 31 Lafayette-class submarines should still be in service with a total of 496 tubes for MIRVed SLBMs on board. In addition, 10 Ohio-class submarines for Trident SLBMs will have been built, with an eleventh to be completed in 1986 but which is here aggregated to the 1985 estimate. With 264

TABLE 3

American and Soviet ICBM Forces: Silo Inventories

Number	At Signature of SAL-1, May 1972	Mid-1976 Estimates		1985 Projections	
		ICBMs	Warheads	Under SAL Rules[b]	Maximal, No SAL Treaty
United States					
Minuteman-1	300[a]	0	0	0	0
Minuteman-2	500	450	450	440[c]	0-450[e]
Minuteman-3	200	550	1,650	560[d]	1,000 + (part mobiles)
Titan-2	54	54	54	0	0
Soviet Union					
SS-7	190	105[f]	105	0	0
SS-8	19	19	19	0	0
SS-11 Mod. 1	970	870[g]	870	9[k]	0-several hundred
SS-11 Mods. 2, 3	0	61[h]	61	61	61
SS-13	60	60	60	0	0
SS-16	0	0	0	60	Mobiles, many
SS-19	0	90	540 }	1,032[l]	1,500[m]
SS-17	0	10	40 }		
SS-9 Mods. 1, 2	288	265[i]	265	0	0
SS-18 Mod. 1	0	20[j]	20	20	20
SS-18 Mod. 2	0	0	0	288	480
Medium and Intermediate Range Ballistic Missiles					
SS-4	500 } + 400 re-loads		500	0	0
SS-5	100 }		100	0	0
SS-20	0		0	Full replacement of SS-4s/SS-5s	

NOTES ON TABLE 3:

(a) The Minuteman-1s (IOC 1962) were being replaced by Minuteman-3s.
(b) Vladivostok rules. See text.
(c) A residual. If none of the MIRVed units are allocated to bombers; if the 31 Poseidon boats retain 16 MIRVed SLBMs each; if 11 Ohio-class submarines will be operational or on sea trials by 1986-1987 with 24 SLBMs each all MIRVed; then out of 1,320 allowable MIRVed units 560 units only will remain for ICBMs. Hence only 10 more Minuteman-3s can be deployed, leaving 440 Minuteman-2s (then 20 years old; but replacements would also have to be single-warhead ICBMs).
(d) As above.
(e) If SAL negotiations break down, additional ICBM deployments would certainly take place, but these would not require the old Minuteman-2 silos.
(f) On the basis of SLBM replacement. "Over 800 SLBMs" are officially estimated in 60 nuclear submarines. Using the latter figure, the SLBM tubes are more precisely estimated at 825 (7 H-class = 21 tubes; 34 Y-class = 544; 11 D-Is = 132; 8 D-IIs = 28). Under SAL-1 rules the 825 SLBM level requires that 85 old ICBMs be dismantled (825 — 740 baseline level). Arbitrarily, it is assumed here that all 85 are SS-7s, leaving 105 in the force.
(g) The 61 silos under construction in 1972 are assumed for late-model SS-11s and not for SS-17s or SS-19s. Of 1,091 total "light" silos allowable, 60 are allocated to SS-13s, 61 as above 870 to SS-11s Mod. 1s, and 100 converted. Of the latter most are known to be SS-19s but the 90/10 split is a low-confidence estimate.
(h) It is possible that not all 61 new light silso are filled as yet.
(i) A residual. Other-type total considered more certain.
(j) The 20 heavy ICBM silos under construction in 1972 cannot have been used for SS-9s (out of production) or SS-18 Mod. 2s (not yet in production).
(k) A residual. Out of 2,400 Vladivostok units 1,320 MIRVed ICBMs and 950 SLBMs are deducted; no bombers are included; 130 units remain. Of these 60 are allocated to solid-fuel ICBMs and 61 to late-model SS-11s, leaving nine.
(l) The 308 sublimit on heavy ICBMs would prevent deployment of more than 288 MIRVed heavies if the SS-18 Mod. 1s are retained. Hence (1,320 — 288) 1,032 units remain for SS-17s and -19s.
(m) High estimate of Soviet production capacity.

SLBM tubes in the Ohio class, SLBMs would account for 760 of the 1,320 allowed MIRVed-missile units. This would leave 560 for land-based ICBMs, only 10 more than now deployed. The reasoning behind the attribution of Soviet MIRVed units to the land-based ICBMs exclusively is of course more speculative: first, although the latest Soviet SLBM is suitable for high-precision modifications, *all* SLBMs are less accurate than ICBMs, other things being equal; Soviet SLBM warheads are larger than those on American SLBMs, but the counterforce intent in Soviet strategic policy should still accord primacy to the inherently more accurate land-based weapons; second, while no fewer than four different Soviet MIRV post-boost vehicles have already been tested (some extensively) on land-based ICBMs, none is known to have been tested on a SLBM. It is true that it would be inconsistent with traditional Soviet practices if at least an experimental MIRVed deployment did not take place at sea. However, should a SAL rule be established whereby *any* MIRVed deployment in a particular type of silo (or in a particular type of submarine) would bring *all* weapons in such housings under the MIRV sublimit, it is possible that Soviet defense planners would sacrifice their usual penchant for experimental deployments.

Here and in Table 3 it has been assumed that *none* of the allowable MIRVed units would be assigned to bomber aircraft fitted with long-range cruise-missile launchers. It does not, at the time of writing, seem probable that the inclusion of such bombers in the MIRVed weapon category could be accepted by the United States Congress. Further, the American air force does not appear to show much enthusiasm for the development of long-range cruise-missiles whose deployment would reduce the role of the bombers to that of flying launching pads. On the other hand, it seems unlikely that the Soviet defense establishment will be able to produce air-launched cruise missiles by 1985. For this very reason Soviet SAL negotiators are expected to struggle mightily to bring long-range cruise missiles and their carriers under the overall weapon ceiling *and* within the MIRVed weapon sublimit. But the era of unilateral concessions and one-sided Russian advantage may be over in the SAL negotiations, and this in turn may mean that there will be no cruise-missile limitation; or it may mean that there be no SAL Treaty at all.

The Lesser Nuclear Powers. Among the lesser nuclear powers, only China is proceeding with a major land-based ballistic missile program. British efforts in this sector ceased more than decade ago with the cancelation of the "Blue Streak" IRBM, and the French land-based missile force consists of only 18 IRBMs (another nine may be added), at present equipped with 150 Kt fission warheads. There are plans for replacing them with MRV thermonuclear warheads in the megaton range. Although housed in protected silos the French IRBMs are inevitably very vulnerable: inaccuracy increases with range in existing ballistic missiles, and Soviet weapons should have *much* higher kill probabilities against French IRBMs than against more distant American ICBM silos. The French missile force unlike the Russian or the American cannot offer the *assurance* of retaliation even in a small way.

The Chinese land-based missile development program has advanced at a rather slower pace than many had expected, especially as far as full-range ICBM weapons are concerned. China remains a poor country, with small islands of technical sophistication and modern industry in a primitive agrarian economy. Moreover, it seems that the Chinese are following a long-term development strategy rather than searching for quick-fix solutions. For example, *large-scale* facilities for producing nuclear materials have been built and substantial resources have been allocated to solid-fuel rocket propulsion (at present Chinese missiles are liquid-fuel vehicles). In the meantime, China deploys perhaps 50 MRBMs with a range of 600 n. miles (enough to reach Soviet cities in eastern Siberia); and a roughly similar number of 1,500 n. mile-range IRBMs. Both weapons are primitive: only a few of the IRBMs are in protected housings, and the MRBMs have a very slow reaction time which compounds their vulnerability.

More capable weapons—a 3,000 n. mile multi-stage ICBM and a very large full-range ICBM—are both under development. Some of the former may be operational, though largely untested; the latter may not be ready for operational deployments until the mid-1980s. It seems reasonable to expect that the Chinese will concentrate their shorter term efforts on the production of limited-range weapons to be targeted on the Soviet Union, rather than full-range ICBMs which could also strike at the United States.

The deployment of a reliable deterrent against Russian nuclear coercion or outright attack would seem to be a most urgent national priority for China, while a strategic force targeted on the United States would serve only second-order political goals. The present Chinese land-based missile force is not only small but also poorly protected against a disarming counterforce strike (which need not necessarily require the use of nuclear weapons—and which thus may not entail the full political costs of a *nuclear* strike to the Soviet Union). By the time a Chinese ICBM force is fully developed, it is possible that *all* fixed-site weapons will have become vulnerable. Anticipating this, the Chinese may therefore proceed directly to deploy *mobile* ICBMs in the mid-1980s.

Submarine-Borne Ballistic Missiles (SLBM)

With fixed-site ICBMs becoming increasingly vulnerable as warhead accuracies improve, the importance of the submarine element in the strategic forces of the nuclear powers has much increased and is expected to increase still further. And it is precisely in this sector of the strategic competition that the shift in the balance of forces between the Soviet Union and the United States has been most dramatic.

Until recently the *qualitative* superiority of the American SLBM force was so great that its numerical superiority—though large also—scarcely seemed to matter. From the beginning, all American missile-submarines have had nuclear propulsion and have been equipped with solid-fuel missiles of great reliability meant for underwater launch. The range of the missiles was 1,200 nautical miles in the first (Polaris A-1) version operational by 1960. Subsequent versions increased the range to 2,500 n. miles by 1964. In contrast, throughout the 1960s the Soviet SLBM force was little more than an improvisation. Some nuclear (H-class) and some diesel-electric (Z-class and G-class), all the submarines were makeshift conversions of torpedo-firing attack types; as a result their missile capacity was limited to three tubes housed in the conning tower as opposed to the 16 tubes in the Polaris hull. Russian submarines were much less quiet in underwater cruise than the Polaris boats and correspondingly more vulnerable to sonar detection. Further, the diesel-electric G-class boats could not stay underwater at their firing stations without using snorkel tubes, and were thus additionally vulnerable to radar detection. The shortcomings of Soviet submarines were compounded by those of their missiles. The G-class boats were originally fitted with SS-N-4 missiles with a maximum range of less than 350 n. miles; this would of course force the launching submarine to approach coastal waters in order to keep the weapon on target. Worse, the SS-N-4 could only be fired from the surface, thus imposing a sea-state limitation on the system, as well as making it yet more vulnerable to detection. The H-class and some G-class boats were equipped with a missile capable of underwater launch, the SS-N-5 which became operational in 1963; this had a maximum range of 700 n. miles.

For these reasons, the numerical advantage of the American SLBM force, understated its vast superiority. Even nonexpert political observers could perceive the net qualitative advantages which put the U.S. submarine force on an altogether higher plane of effectiveness.

The United States did not, however, preserve this margin of advantage. Innovation was limited to qualitiative improvements in the SLBMs—so long as these were compatible with the tube dimensions of the Polaris boats. Nevertheless, by 1971 with the introduction of the Poseidon C-3, a major increase in effectiveness had been achieved. Though its maximum range was no greater than the previous Polaris A3 at 2,500 n. miles, the Poseidon was fitted with a MIRV system for up to 14 warheads of 40 kilotons (with a range tradeoff). The normal loading appears to be limited to 10 of these small warheads; even so, against cities, industrial centers, airfields, and other unprotected targets, each Poseidon can be up to 10 times as effective as the single-warhead megaton-range Polaris A2s, and more than three times as effective as the triple-warhead (but not separately targetable) Polaris A3. The United States did not, however, add the Poseidons to the submarine force. Instead, existing Polaris boats were converted to accommodate the new missile, and the self-imposed 41-boat limit on the force, established by 1964 and attained in 1967, was maintained.

American restraint on the one hand and a very rapid Soviet deployment program on the other enabled the Soviet Union to reverse the numerical superiority of the American SLBM force in the course of a few years, while at the same time reducing very considerably its qualitative advantage. Starting in 1968, a much-improved nuclear-powered submarine, the 16-tube Y-class, was deployed at so rapid a rate that 34 of these 16-tube boats were in service by 1975. Y-class boats were fitted with an underwater-launch SS-N-6 missile whose range of 1,300 n. miles, while still inferior to that of the latest American weapons, was a very considerable improvement on the earlier Russian SLBMs.

By 1973, a new Soviet SLBM, altogether larger than the SS-N-6, was deployed, the SS-N-8. This weapon has been tested at ranges of 4,200 n. miles (that is, 70 percent more than the Poseidon range). At first SS-N-8s were deployed on board 12-tube D-1 submarines, essentially similar to the Y-class but stretched to accommodate the much larger launch tubes required by the new missile. By 1975, a new and much larger submarine, the D-2 was observed undergoing sea-trials. The D-2, 500 feet long and with a surface displacement of well over 8,000 tons, is fitted with 16 of the large missile tubes needed to house the SS-N-8. Concurrently, in the characteristic Soviet deployment style, the SLBMs sized for the smaller Y-class tubes are also being upgraded in range and warhead effectiveness: the latest Mod. 3 of the SS-N-6 has a MRV (not MIRV) warhead system and a range of 1,600 n. miles.

The first 10 Polaris boats, of the George Washington and Ethan Allan classes, were built in crash-program conditions and have now been operated intensively for a decade and a half. Partly to replace these boats and partly to accommodate a much larger SLBM, a new class of American missile submarines is now under construction: the Ohio-class, of wholly unprecedented size for submarines with a submerged displacement of 18,700 tons. They are to be fitted with 24 large-diameter missile tubes which *could* accommodate an ICBM-sized missile giving full intercontinental range even with a large throw-weight. At present, however, no such missile is being developed. Instead the first of 11 planned Ohio-class ships are to go in service in 1979-1980 with the interim Trident C-4 missile whose diameter is compatible with the Poseidon missile tubes. It is envisaged that these new SLBMs will be deployed on board 10 of the Lafayette-class Polaris boats which went into service between 1963 and 1967. The Ohios are likely to exceed in capability their Soviet counterparts in the 1980s. It should be noted, however, that the Trident C-4 missile (which will not become available until 1979) has already been exceeded in range by the SS-N-8, a weapon which has been in service since 1973.

It is clear that the pace of Soviet and American SLBM deployments has scarcely been hampered by the SALT 1 accords. The latter set final ceilings on "modern" SLBM tubes of 950 units for the Soviet Union as against 710 units for the United States; concurrently the number of nuclear submarines was limited to 62 boats for the Soviet Union and 44 boats for the United States. It will be recalled that even those observers who welcomed the SALT-1 accords, and who saw the ICBM force-level imbalance as not unreasonable in the light of American superiorities elsewhere, were nevertheless disturbed by the SLBM provisions. Aside from the gross and highly visible inequality of the force ceilings, the exchange provisions were also highly favorable to the Soviet Union. No more than 25 Y-class (400 tubes) and eight H-class (24 tubes) nuclear boats were operational at the time, but in the SALT-1 Protocol it was stated that there were 740 Soviet SLBMs on board *nuclear* submarines operational or "under construction." The practical consequence of this was that the Soviet Union was given 84 more baseline units than the United States. The Soviet Union was given the option of adding 210 SLBMs to the baseline force in exchange for the withdrawal of older SLBMs and ICBMs. As against this, the United States could add only 54 tubes, by scrapping the Titan-2 force.

The complex and ambiguous SLBM provisions of the SALT-1 accords are to remain in force until October 1977; their practical consequence is therefore small: the Soviet Union could hardly exceed the ceilings anyway while no new American submarines will come into service until 1979 at the earliest. On the other hand, in anticipation of an extension of the constraints, the SAL-1 ceilings on "modern nuclear-powered submarines" (44 for the United States and 62 for the Soviet Union) may well have influenced Soviet design decisions: with 34 Y-class boats already built, only 28 more hulls were allowed; if all were 12-tube D-1 designs, this would mean that Soviet Union could deploy only 880 SLBM tubes in all. To reach the full allowance of 950 SLBMs a submarine with more tubes than the D-1 was needed, and the new D-2 ship fitted with 16 tubes fulfills this need. Of course, independently of the SAL-1 constraint, the D-2 is a more economical platform since the overhead cost for each missile must be significantly smaller.

Under the terms of the Vladivostok agreement-in-principle, up to 2,400 (or perhaps 2,200) strategic weapons may be deployed by each side, including up to 1,320 delivery systems fitted with MIRV. It is not clear how these units will be allocated on each side. As far as the United States is concerned, the present Ohio-class submarine program calls for the deployment of 11 ships with a total of 264 tubes. If these were added to the 496 Poseidon tubes, SLBMs would account for all but 560 of the 1,320 units in the MIRVed missile sublimit. By 1982 the first 10 Polaris boats (160 tubes) will all have reached their twentieth year of continuous operation,

TABLE 4
Salient Characteristics of American and Soviet Ballistic-Missile Submarines/SLBMs

I. SUBMARINES

	Year IOC	Propulsion	SLBM Tubes	Length	Estimated Surface Displacement
United States					
George Washington	1960	nuclear	16	381.5 feet	5,900 tons
Ethan Allan	1961	nuclear	16	410.5 feet	6,900 tons
Lafayette	1963	nuclear	16	425 feet	7,320 tons
Ohio	1979/80	nuclear	24	560 feet	15,000 tons
Soviet Union					
G-class	1960	diesel-elec.	3	320 feet	2,300 tons
H-class	1964	nuclear	3	380 feet	3,700 tons
Y-class	1968	nuclear	16	428 feet	8,000 tons
D-1 class	1973	nuclear	12	450 feet	8,500 tons**
D-2 class	1976/77	nuclear	16	500 feet	9,000 tons**

II. SUBMARINE-BORNE BALLISTIC MISSILES

	Year IOC	Status*	Max. Range (n.miles)	Warhead System
United States				
Polaris A-3	1964	OL	2,500	MRV : three warheads of cca. 200 Kt.
Poseidon C-3	1971	O	2,500	MIRV: (std.): ten of cca. 40 Kt.
Trident C-4	1979	R & D	4,000+	MIRV: (std.): ten of cca. 100 Kt.
Soviet Union				
SS-N-4	1960	OL	350–	Single, Megaton range
SS-N-5	1963	OL	700–	Single, Megaton range
SS-N-6 Mod. 1	1968	O	1,300	Single, Megaton range
SS-N-6 Mod. 2	1973	OL	1,600	Single, Megaton range
SS-N-6 Mod. 3	1975	OL	1,600	MRV: three warheads of cca. ½ Mt.**
SS-N-8	1973	O	4,200	Single, MRV (not MIRV) likely

*Status: O = operational; OL = limited deployment; R & D = not yet in production.
**Low-confidence estimate.

and it is the present position of the Navy that this is the rough limit on their useful life. On the other hand, the increasing threat to land-based ICBMs and the high costs of land-mobile systems may prompt a partial shift in the strategic arsenal to SLBMs (and/or cruise missiles), and this could mean that as more Tridents are deployed some Poseidon boats will be converted back to non-MIRV missiles. A third and attractive alternative would be the deployment of a smaller, slower, and *much* cheaper submarine than the fast 18,700 ton Ohios.

In the meantime, the only near-term change in the U.S. force will be the completion of the Polaris A3-to-Poseidon conversion program in the Lafayette-class submarines. Thus far, 27 have been converted; the other four are now in conversion, and the last of the 31 boats is due to enter service in 1978. Under present plans 10 of these submarine are to be fitted with Trident missiles, with conversion beginning in 1979.

The 950-unit SLBM force-level set by the SALT-1 agreements opened a wide range of deployment options to the Soviet Union, though the 62-unit limit on the number of modern submarines would evidently become an operative constraint. No more Y-class boats are likely to be built beyond the 34 now operational (though their missiles will no doubt continue to be upgraded). It is certain that the Soviet Union will exercise its full exchange option under Article III of the Interim Agreement, replacing older ICBMs and SLBMs with modern SLBMs, but it is not yet clear what mixture of 12-tube D-1 and 16-tube D-2 boats will be deployed. If the Soviet Union's goal is to maximize the number of boats (to reduce the overall vulnerability of the force) it would build only D-1s; but a maximum force of 34 Y-class and 28 D-1 class boats would only accommodate 880 tubes; this would still leave 70 SLBM tube units unclaimed. If the Soviet goal is to maximize the number of SLBMs at the lowest overhead cost, then only D-2 boats would be added to the force. (A reasoned projection is presented in Table 5.)

It seems unlikely that the Russian Navy will be allocated significantly more than 950 of the allowable weapon units under a Vladivostok-style agreement. Even if the Soviet bomber force is fully excluded (which seems unlikely), the remaining units under a 2,400 unit or 2,200 unit ceiling would barely justify the very great investment that the Soviet Union has made in its four different new ICBMs. For the reasons given above, it seems unlikely that *any* of the MIRVed units would be allocated to the SLBM.

It seems probable that by 1977 or soon thereafter the seven H-class boats, vulnerable and grossly uneconomical platforms, will be withdrawn (or converted back into torpedo-firing "attack" boats). It would also be

TABLE 5

American and Soviet SLBM Forces: Submarines and SLBM Tubes in Inventory

	At Signature of SAL-1, May 1972	Mid-1976 Estimates		1985 Projection	
		Boats/SLBMs	Warheads	Under SAL Rules	Maximal
United States					
George W./Ethan A. Submarines	10	10	n/a	0	0-10[a]
Polaris A-2 Missiles	160	0	0	0	0
Polaris A-3 Missiles	0	160	160	0	0-160
Lafayette-class Submarines	31	31	n/a	21	14
Polaris A-3 Missiles	432	0	0	0	0
Poseidon C-3 Missiles	64	496	4,960	336[b]	224[c]
Ohio-class Submarines	0	0	n/a	10[d]	11-13[e]
Trident C-4 Missiles	0	0	0	240	264-312
Conv. Lafayette Submarines	0	0	n/a	10	17
Trident C-4 Missiles	0	0	0	160	272
Totals: Boats (inc. in conv.)	41	41	–	41	42-44
SLBMs (do.)	656	656	–	736	760-968
Warheads: S/MRV	592	–	160	0	160
MIRV: 40 Kt.	640	–	4,960	3,360	2,240
MIRV: 100 Kt.	0	–	0	4,000	5,360-5,840
Soviet Union					
G-class Submarines	21	20	n/a	–[f]	–[f]
SS-N-4 Missiles	30	27	27	–	–
SS-N-5 Missiles	33	33	33	–	–
H-class Submarines	8	7	n/a	0	–[g]
SS-N-5 Missiles	24	21	21	0	–
Y-class Submarines	25	34	n/a	34	34
SS-N-6 Missiles	400	544[h]	544	544	544
D-1 class Submarines	0	11[i]	n/a	11	11
D-2 class Submarines	0	8	n/a	17	40[j]
SS-N-8 Missiles (or follow-on)	0	260	260	404	772
Totals: Boats (nuclear boats)	54 (33)	80 (60)	n/a	(62)	(85)
SLBMs: single RV or MRV	487	885	885	948	1,316
Of which "modern"	400	804[h]	804	948	1,316
Of which MIRV	0	0	0	0	772[k]

NOTES ON TABLE 5:

SAL-1 Rules on SLBM/submarine deployments: all post-1964 SLBMs on any type of boat are counted; all SLBMs of any type on nuclear boats are counted.

(a) These boats might remain in service but not for much longer.
(b) Out of 496 tubes, 160 are to be converted to the C-4 missile, leaving 336 still fitted with Poseidon C-3s.
(c) Assumes three conversions p.a. beyond 1983.
(d) Eleven boats are now funded; last to be completed by 1986.
(e) Assumes a 1,2,1,2, production rate, with extra boats funded in FY 1978 and FY 1980.
(f) These decrepit boats may yet remain in service.
(g) It is conceivable that the 7 H-class boats would remain in service even though their hulls could be converted into more useful attack boats (N-class type).
(h) Some additional SS-N-6s are fitted on G-class boats.
(i) Eleven D-1 boats are known to be operational. "Over 800 SLBMs" are officially estimated, as well as "about 60" nuclear submarines; hence 8 additional boats are presumed beyond 34 Ys, 7 Hs, and 11 known D-1s. If they are D-2s, it comes to 825, which is closer to the "over 800" estimate.
(j) Assumes a production rate of four p.a. post 1980.
(k) It is assumed that no Y-class compatible MIRVed SLBM is deployed; that Y-class not converted to anti-shipping use with SS-N-13s; that only SS-N-8s deployed in MIRVed version.

reasonable to assume that the diesel-electric G-class boats will be withdrawn, too. On the other hand, the Soviet Union has in the past shown a consistent tendency to retain major weapons in service well past the point of *technical* obsolescence.

The Lesser Nuclear Powers. Both Britain and France have followed the superpowers in deploying SLBMs on board nuclear-powered submarines, albeit in small numbers. But while Britain received both complete missiles (Polaris A3s) and other technical aid from the United States, the French program was largely based on domestic technology. This has resulted in the slower deployment of less capable forces, and at considerably higher cost than the trouble-free British program. The first British Polaris boat became operational in March 1968 and the full force of four boats, with a total of 64 tubes for triple-warhead Polaris A3 missiles is now in service (with two boats normally at sea at any one time). The French submarines also have 16 tubes each and under current plans, a total of five (and perhaps six) boats are to be deployed. At present three boats are in service, two with (M-1) 1200 n. mile missiles fitted with 500 kiloton single warheads, and one with (M-2) 1600 n. miles missiles. A new (M-20) missile with a megaton-yield warhead was deployed in a fourth submarine during 1976.

China has had a single G-class boat for many years but no missiles; a nuclear-powered submarine is being built, but it is not known if it is a SLBM carrier or an attack boat (the latter being more probable). So far Chinese strategic missiles have all been based on liquid-fuel but it is known that solid-fuel rocket technology, essential for effective SLBMs, is being developed. It is possible of course that the Chinese may take a different approach to the problem of fixed-site missile vulnerability, by developing *land*-mobile ballistic missiles. In Chinese conditions, a road-mobile system for IRMBs or even ICBMs would be quite feasible, and it would obviously be much cheaper to rotate IRBMs or ICBMs on moving wheeled platforms (with or without shelter-stops) than to keep them at sea, inside very expensive nuclear-powered submarines.

Manned Long-Range Bombers

Stringently optimized to carry out a single and irrevocable nuclear strike, ICBMs and SLBMs are devoid of the flexibility which is a normal attribute of military forces. They can neither deliver a varied weapon load, nor visibly heighten their state of alert, nor carry out non-nuclear missions. The tight limits on the *political* applicability of strategic-nuclear force have an exact corollary in the technical-operational rigidity of missile forces. Manned long-range bombers are by contrast very flexible weapons: they may be used for strategic or battlefield bombardment, nuclear or non-nuclear; they may be deployed routinely, or kept on ground alert or rotated in airborne alert; they can be recalled before reaching their destination; they can provide some real-time information on the effects of their own strikes, and in general can bring force to bear anywhere on the planet, including the many places where there is no tactical air power deployed within range. Bombers are here evaluated exclusively in strategic-nuclear terms, but in estimating their overall utility it is necessary to include the full range of their capabilities, including non-strategic (and non-nuclear) capabilities.

Manned bombers are the only category of strategic weapons in which the United States still retains a net superiority both qualitative and quantitative. This, in spite of the fact that the principal instrument of this superiority is an aircraft which embodies design concepts now a quarter of a century old, and which has not been in production since October 1962. But the B.52 *Stratofortress,* owing to its sound basic structure and large payload, has proved to be an exceptionally versatile aircraft: originally designed for the high-altitude delivery of heavy gravity bombs (at a time when thermonuclear devices were heavy indeed), it has now become through successive modifications a low-level bomber primarily armed with supersonic air-ground missiles. Repeatedly modernized since first introduced in 1955, built with major improvements in the "G" and "H" models (first deployed in 1959), and reequipped with new weapons and electronics more than once, the B.52 cannot nevertheless be converted into a supersonic aircraft, or one capable of extreme low-level "contour flight"—that is, the sort of aircraft which is designed specifically to cope with modern air defenses. The USAF did deploy a supersonic bomber during the 1960s, the B.58 Hustler; and the FB-111, which is both supersonic and contour-flying, is now in service. But neither aircraft has been produced in quantity to replace the B.52s: for all their special attributes, they lacked the sheer load capacity of the B.52s whose take-off weight of 400,000 pounds or more accommodates enough fuel for a two-way intercontinental range, and a large weapon and countermeasure payload.

Of the 744 B.52s built, 348 remain in the active inventory; of these in turn 316 are deployed operationally and rotated in ground alert. The 151 "G" and 90 "H" model aircraft in the operational force are equipped to carry the SRAM stand-off missile, which can be used both as a "percursor" to destroy air defenses, and to attack primary targets up to 120 miles beyond the launching aircraft (gravity bombs are also carried).

The key to bomber pentration against modern air defenses is neither speed nor altitude, since anti-aircraft missiles can easily prevail in both; instead enemy radars on the ground must be evaded by low-level flight and/or neutralized by jamming and/or deceived by means of electronic countermeasures (ECM). Hence the load capacity requirement, since low-level flight is far less economical than high-altitude cruise, while ECM needs require significant payloads. In spite of the B.52's great range (up to 12,500 miles unrefueled in the "H" model) it was therefore found necessary to deploy tankers along with the bombers. Thus 657 KC-135 tanker aircraft remain in the active inventory to support B.52 operations and to give the much smaller FB-111s an intercontinental reach as well as to serve all other USAF refuelling needs.

Sixty-six FB-111s are deployed operationally. Unlike the B.52s, these are low-altitude penetrators, intended to cross heavily defended areas by flying only a few hundred feet off the ground. The small ECM payload of these small aircraft is not a significant operational limitation at present, since the high-speed contour-flying FB-111 is not vulnerable to present Soviet air defenses. On the other hand, the small weapon load of these aircraft means that if all USAF bomber targets were assigned to FB-111s, a much larger and much more costly bomber force would be needed than the one presently maintained.

The B.1 program is intended to produce an aircraft which combines the economical payload capacity of the B.52s (24 SRAMs) with the flight performance of the FB-111. A total of 244 B.1s are planned for a procurement cost in excess of twelve billion FY 1976 dollars. All these aircraft would be operational by 1985, with the first coming into service by 1979, but at the time of writing no final decision has been made in Congress, and the future of this program remains in doubt. There is also an operational uncertainty: the "underflight" of ground-based radar is a tactic which can itself be defeated by "look-down" airborne radars interconnected with the command of fighter-interceptors themselves equipped with "snap-down" missiles. Both the United States and the Soviet Union have developed large-capacity airborne "look down" radars; the American (AWACS) is known to be effective; the Soviet (Moss) is thought to work only over water. But it is possible that a Soviet system comparable to AWACS will be deployed by the mid-1980s. If so, "underflight" would no longer offer a high probability of successful penetration though very low-level contour-flight modes might still do so.

The Soviet air defenses which American bombers have to contend with are very extensive, comprising some 2,600 interceptors (half being highly capable heavy-weight fighters with large radars), 12,000 anti-aircraft missiles, and over 5,000 radars. Nevertheless, these forces cannot preclude bomber penetration; however, they have forced the United States to deploy a sophisticated bomber force, and to accept the high cost of *low-level* penetration. American air defenses have by contrast been reduced to a level which is not much more than symbolic. A total of only 61 radar warning units remain in service, all strategic SAMs have been withdrawn, and the manned interceptors controlled by the joint U.S.-Canadian air defense command (NORAD) number only some 360, not enough to defend the vast air perimeter, and barely enough to check out unauthorized intruders.

Acting on the principle that the self-imposed lack of ballistic-missile defenses would make anti-aircraft defense aleatory, the United States gradually withdrew the vast air defenses deployed during the late fifties and early sixties. Acting on the principle that even a partial defense is better than none, the Soviet Union by contrast invests vast resources in maintaining and upgrading its air defenses. The practical consequences of this is that almost *any* aircraft with the requisite range could in fact reach targets in the United States, without requiring the low-level flight and countermeasure capabilities of the American bomber force. It is in this context that the strategic worth of Soviet bombers must be evaluated: inferior though their quality might be, as against American air defenses they have much greater capabilities than American aircraft with these same specifications would have against Soviet air defenses.

During the 1960s, the bomber force was obviously treated as distinctly secondary in the Soviet strategic arsenal. Both a full jet and a turboprop

TABLE 6

American and Soviet Manned Long-Range Bombers: Salient Features and Force Levels

	Year IOC	Gross Weight Max. Take-off	Max. Range* Int. Fuel	Max. Weapon** Load	Number Deployed in Strategic Roles			
							1985	
					May 1972	Mid-1976	SAL***	No SAL
United States								
B.52 D	1956	450,000 lb.	9,000 st.m.	60,000 lb.	150	75	75	75
B.52 G	1959	488,000	10,000	75,000	165	151	151 [a]	151 [a]
B.52 H	1959	488,000	12,500	75,000	90	90	90(b)	90(b)
[B.1 [c]	1980	395,000	6,100+	75,000+	0	0	244	244]
FB-111	1969	114,000	3,800	37,500	66	66	66	66
Soviet Union								
Tu-95 (Tu-20) Bear	1956	360,000	8,000	40,000	100	105	0	100?
Mya-4 Bison	1956	250,000	6,000	20,000	40	35	0	35?
Tu-26 (?) Backfire	1975	280,000	5,000	20,000	0	40 [d]	0-300+ [e]	300+
Tu-16 Badger	1955	150,000	4,000	20,000	500	475	0 [f]	0 [f]
Tu-22 Blinder	1962	175,000	1,500	12,000	200	175	175	1,975

*This is the range in optimum conditions. Operational ranges with weapon loads and low-level flight profiles are much shorter.

**Theoretical maxima, with corresponding range reduction.

***Vladivostok rules.

(a) Number of operational (UE) B.52 Gs may decline further; only 21 active inventory units are available for attrition.

(b) As above; 6 units available.

(c) Program not finalized.

(d) 50 Backfires (Bs) have been built; 10 are assumed in naval service.

(e) Calculated on the basis of (i) estimated production rates, (ii) Tu-16-like allocation between SAF and naval use. 0-300 = depends on how SAL rules are formulated.

(f) These aircraft may be retained in service, in spite of their low quality.

bomber were deployed operationally after 1955 but neither was produced in large numbers. Out of an inventory of some 265 aircraft of both types, some 35 full-jet Mya-4s (Bison) and 105 turbo-prop Tu-95s (Bear) are thought to be deployed in a strategic strike role; the remainder have been converted into tanker and reconnaissance aircraft. Neither aircraft has modern stand-off missile or powerful countermeasure equipment, but then again neither would need it.

After a period of many years in which it seemed that the Soviet Union had simply abandoned the heavy bomber as a strategic instrument, the variable-geometry supersonic Backfire was introduced operationally in 1975. Two thirds of the size of the B.1, this aircraft is estimated to have an unrefueled range on the order of 5,000 miles. Able to fly at high altitudes in the absence of effective American air defenses, and using its refuelling equipment, the Backfire would be perfectly capable of two-way strategic delivery.

It has been reported that in the SAL negotiations the Soviet Union has insisted that the Backfire is not a "strategic weapon" and should not be counted against the overall weapon ceiling. On the other hand, the Soviet Union insists that American heavy bombers *should* count against the ceiling, and that bombers equipped with long-range cruise missiles (600 KM+) should also count against the MIRVed launcher sublimit of 1,320. It has further been suggested that the United States might accept Soviet deployment guarantees and agree to exclude Backfire aircraft from the limitations. There is no means of verification which can preclude the swift redeployment of Backfire aircraft at forward bases; nor can the conversion of large transports into tankers be precluded. Should the Soviet position be accepted, a new and much less reliable standard of verifiability would therefore be legitimized in the SAL process.

If Backfires are included in the SAL count it seems reasonable to expect that the Soviet Union would not allocate any allowable weapon units to manned bombers (deploying instead 2,400 or 2,200 SLBMs and ICBMs). In any case the Soviet leadership would then have the unenviable task of arbitrating the bureaucratic battles of the Navy and Strategic Rocket Force on the one hand, and the Soviet Air Force on the other. Certainly if effectiveness calculations intrude in the decision, and if American airspace continues to remain virtually undefended, a force of Backfires should have a weighty claim to deployment in lieu of some of the more marginal ICBMs and SLBMs.

By contrast the American bomber force will certainly account for a significant proportion of any overall weapon ceiling (though not of the MIRVed weapon sublimit). Even if the B.1 is not built, 250 or so B.52s

TABLE 7

American and Soviet Manned Long-Range Bombers: Estimated Weapon Loadings (Strategic-nuclear mission)

	Maximum Loadings	Standard Loads	Type Totals (standard)
United States			
B.52 D	8+ Laydown Mk.28 or B.61 bombs[a]	4 bombs	300 bombs
B.52 G and H	4 bombs; 20 SRAM missiles[b]	4 bombs; 6 SRAMs	964 bombs; 1,008 SRAMs[c]
(B.1)		24 SRAMs/ALCMs	5,280 SRAMs/ALCMs[d]
FB-111	6 SRAMs	2 SRAMs[e]	132 SRAMs
TOTAL: Mid-1976 Estimate:	1,140 SRAMs; 146 Hound Dogs[f]; 1,264 bombs.		
1985 Estimate:	7,208 SRAMs/ALCMs[g]; 132 SRAMS; 1,264 bombs.		
Soviet Union[h]			
Tu-95 (-20) Bear	1 AS-3; 4 bombs	1 AS-3; 1 bomb	105 AS-3s; 105 bombs
Mya-4 Bison	4 bombs	2 bombs	70 bombs
(Backfire)[i]	(Mod. ASMs)		80-160 bombs[i]
TOTAL: Mid-1976 Estimate:	105 ASMs (AS-3s); 255-355 bombs.		
1985 Estimate:	600 ASMs; 600 bombs		

(a) Laydown (= gravity = "iron") bombs; Mk.28 fixed-yield 5 Mt. (?); B.61 variable yield; std. 1 Mt. (?).
(b) Short Range Attack Missile; warhead estimated at 200 Kt.
(c) A total of 1,040 SRAMs are deployed as unit equipments; of these 132 are assumed in service with FB-111 force. The 73 B.52 G/Hs unequipped with SRAMs are assumed armed with the older Hound Dog long-range ASMs with megaton yield warheads (308 are available).
(d) Air Launched Cruise Missiles: Long-range high accuracy ASMs with small-yield warheads. Would be in service in 1980s unless constrained by SAL rules. (Shorter-range ALCMs may still be deployed.)
(e) FB-111s can carry two SRAMs internally. External loads seriously degrade FB-111 performance.
(f) See (c).
(g) Assumes procurement of additional SRAMs and/or ALCMs to use fully B.52 rotary launcher capacity (8 SRAMs or ALCMs).
(h) All low-confidence estimates.
(i) If Backfire is deployed in a strategic strike role, it is reasonable to expect that a modern ASM will be deployed also; at present, if operationally deployed in a strike role, the Backfires would carry several gravity bombs.

will remain in service throughout the 1980s. If the B.1 program does proceed as planned, then almost 500 allowable weapon units will be allocated to bombers, along with 1,000 ICBM and 736 SLBM units. (This total force would still be below the 2,400 unit ceiling which was sized for Soviet, and not American, strategic forces.)

In addition to its intercontinental-range bombers, the Soviet Union also deploys large numbers of Tu-16 'medium' bombers with a combat radius of up to 2,000 miles. The Tu-16s, roughly 450 of which are in Air Force service (another 250 are Naval), do not have an assigned strategic role against the United States but they could certainly reach many parts of the United States on one-way missions; it is thought, however, their primary targets are in Western Europe and China.

The Lesser Nuclear Powers. During the 1950s Britain invested heavily in an ambitious bomber program based on no fewer than three different medium-range aircraft models, the so-called V-bombers. Now, however, all but 50 of these aircraft have been withdrawn from service or converted to other uses, and the remaining force of Vulcan B2s has been reassigned to a tactical mission. In theory, therefore, the British no longer have a strategic bomber force. But Britain does retain a large stock of nuclear weapons, and its remaining Vulcan B2s, Buccaneer S2 light bombers (and even F-4M Phantoms) deployed in Germany could all reach targets in the Soviet Union. In fact the Buccaneers have a range similar to that of the Mirage IV, the official "strategic" bomber in French service.

The bomber force was the first element in the French mini-triad to enter service in 1964; it now consists of 52 Mirage IVs of which 36 are deployed on continuous ground alert each carrying a single 70 Kt fission bomb. The Mirage IVs require mid-air refueling to reach targets such as Moscow on two-way missions; 11 U.S.-built KC-135F tanker aircraft are available for this task. Owing to their small payload, the French bombers lack adequate electronic warfare capabilities, while their small fuel capacity prevents sustained low-level flight at high speeds.

So far the Chinese have not flight-tested a strategic bomber of local design, but a force of 63 locally-built TU-16 copies is in service, and these aircraft could in theory reach Moscow on one-way missions even from deep inland airfields in China. Whether they could penetrate Soviet air defenses is another matter. The Russians must calculate, however, that if an all-out attack were launched, a Chinese aircraft or two could reach a major Soviet city. Chinese Tu-16s, as also many tactical aircraft, have a much more realistic attack capability against targets east of the Urals. Although numerically small and technically obsolescent by superpower standards, the Chinese bomber force is important, since the only other Chinese strategic delivery vehicles, the land-based missiles, remain highly vulnerable to attack.

In the case of India and other potential third-level nuclear powers, manned aircraft are likely to be the most easily available means of nuclear delivery. Against large, soft (city-type) targets not protected by sophisticated air defenses, the nuclear delivery mission is not particularly exacting and may be executed even by ordinary fighter-bombers so long as they have reasonable payloads. India for example could use old Canberra light bombers (3,800 mile range) or Su-7 fighter-bombers (900 mile range) to reach targets in Pakistan or Western China; present Indian aircraft would not, however, offer a realistic strike capability against Peking and the rest of northeast China. Of course, no improvised force of aircraft can do more than offer *some possibility* of delivering nuclear weapons against poorly protected targets. A more reliable capability requires specialized aircraft and training, as well as much supporting equipment.

Other Strategic Weapons, Old and New

The ballistic missile launched from a land silo or from a submarine and the long-range manned bomber are not the only instruments of strategic power. Nor is the future evolution of strategic arms constrained by the "triad" pattern which originated in the United States and which all other nuclear powers have tried to imitate in full or in part.

If a weapon *can* reach deep into enemy territory to strike at "homeland" population centers, it is in effect "strategic" whether or not it has full intercontinental range. Weapons in this class include the Russian medium and intermediate-range ballistic missiles targeted on Western Europe and China (see Tables 2 and 3). From the American point of view these weapons are not "strategic" in a formal sense, but if the Soviet Union did not have MRBMs and IRBMs it would have to allocate ICBMs for the purpose. (It is estimated that some 100 SS-11 ICBMs are in fact deployed in MBRM/IRBM fields.) For this reason the deployment of additional MRBMs or IRBMs may be tantamount to an increase in the Russian ICBM force, to the extent that some ICBMs could then be released for use against American targets.

Another group of non-intercontinental weapons with an inherent strategic potential are the American fighter-bombers—deployed in Britain, Germany, and South Korea, as well as the carrier-based attack aircraft deployed with the Sixth Fleet in the Mediterranean and with the Seventh Fleet in the Pacific—all of which could in theory reach targets within the

Soviet Union. These aircraft are the so-called FBS (forward-based systems) on which there has been much discussion in conjunction with the SAL negotiations. It appears that the Soviet Union has insistently argued that these are strategic weapons, fully equivalent to the intercontinental weapons in the official strategic arsenal. But of course the fundamental difference between these aircraft in theatre deployment and the strategic bombers is not in the range differential but rather in their respective vulnerabilities to Soviet attacks and to Soviet air defenses. None of the forward-based aircraft can attain the degree of survivability of the bombers deployed in the United States: the latter are kept on a ground-alert status which corresponds to the length of pre-attack warning, while the FBS cannot count on adequate warning against MRBM/IRBM attack. Moreover, only the 72 heavyweight F-111E strike fighters deployed in Britain have a realistic penetration capability against Soviet air defenses; all the other tactical aircraft lack the fuel capacity for sustained low-level flight. It is true that the A-6s and A-7s on board the carriers have a back-up role in strategic attack plans, but ever since the withdrawal of the A-5 Vigilante light bombers no Navy aircraft have been deployed with a primary strategic strike mission. It is difficult to believe that the Soviet Union has perceived a serious incremental threat in the FBS. More likely the issue was raised because of its political significance: the FBS are a critical link between the NATO alliance and American strategic deterrence, and if the FBS were withdrawn the Alliance would be seriously disrupted.

Russian long-range "cruise" (non-ballistic) missiles deployed on board submarines are in a somewhat similar category. Twenty-seven E-II class nuclear-powered boats, each fitted with eight 550 n. mile-range SS-N-3 missiles, are deployed at present; in addition, the Soviet Navy also deploys 16 J-class diesel-electric submarines each armed with four SS-N-3 missiles. (There are also some obsolescent W-class boats with the same missile.) No less than the American FBS aircraft, these cruise missiles could reach American (coastal) targets with the nuclear warheads. Nevertheless these weapons have not been considered to be "strategic"; their operational characteristics make them much more suitable for anti-shipping missions.

The SS-N-3 cruise missile is in effect a pilotless aircraft, turbojet-powered and guided by radio command. The United States too has experimented with such cruise missiles, deploying the Regulus-I during the 1950s and developing the Regulus-II, both submarine-launched. Until very recently there seemed little point in deploying cruise missiles: they combine the irrevocability and inflexibility of the ballistic missile with the slow flight of manned aircraft; even the size advantage of any one-way automatic vehicle is offset by the guidance problem. But the new-style cruise missiles developed by the United States of late offer radically expanded possibilities: while the Regulus-II weighed 22,000 pounds for a range of 1,000 n. miles, present technology makes it possible to build 2,000 n. mile cruise missiles weighing less than 4,000 pounds. Of torpedo size, instead of the (small) fighter-aircraft size of the Regulus-II, modern U.S. cruise missiles are also much more adaptable to different launch platforms: they may be deployed alongside torpedos in attack submarines, or (in larger numbers) in SLBM boats suitably converted; they can be launched from aircraft (the USAF is developing an air-launched version sized to fit the SRAM rotary launcher), and they could even be forward-based in Europe on small truck-type mobile launchers. The versatility of the weapon derives from its small weight and compact shape (with snap-out control surfaces); these inherently low-cost weapons can therefore be deployed on low-cost and yet highly survivable platforms.

In addition to their other shortcomings, old-style cruise missiles (as still deployed by the Soviet Union) had an inherent guidance problem: if inertially guided (as ballistic missiles are) then accuracies would be very poor, owing to the long time of flight; if guided by command or with external reference, then they would be highly vulnerable to electronic countermeasures. By contrast, present U.S. technology for cruise missile guidance is based on a terrain reference technique (TERCOM) which cannot be interfered with, and which could provide *extreme* accuracies—much higher than those of any ballistic missile now deployed.

The new-style cruise missiles with their small frontal radar cross-sections (comparable to that of a seagull) and extreme low-altitude flight profiles are immune to conventional air defenses. In practice, an effective defense against cruise missiles would require high-quality, high-cost SAM systems around each defended point. While technically feasible, a nation-wide defense of this kind would be hopelessly costly. AWACS-type "look down" systems may offer an alternative to high-quality terminal SAM barriers, but the economics of such an airborne defense do not look promising either. Even if the problem of reliable detection can be overcome against such small targets so close to the ground, interception would still require high-quality (and high-cost) "snap-down" missiles, launched from fighter-type aircraft with sophisticated electronics. In practice, this may mean the use of half-million dollar ASMs against one to two million dollar cruise missiles; when enough interceptors and ASMs are provided to cover all the alternative paths of attack, the cost exchange between cruise missiles and such air defenses will be sharply in favor of the former.

In theory, cruise missiles could make excellent counterforce weapons, because of their extreme accuracy potential; but the fuel economy of their lightweight turbofans (which have made the innovation possible) is not compatible with very high speeds and present U.S. cruise missiles fly subsonically. Thus even with forward deployments, the long time of flight of these weapons would preclude their use against ICBM silos. Nevertheless, the large-scale deployment of cruise missiles could have a revolutionary impact on the strategic relationship between the United States and the Soviet Union. Militarily, the deployment of cruise missiles in place of the Minuteman force would deprive the Soviet ICBM force of its prime counterforce target; at the same time, the large-scale deployment of these highly effective and distinctly "modern" weapons should negate the perceptual-political benefits the Soviet Union may otherwise draw from its very great investment in new ICBMs.

However the inherent advantages of the new cruise missile technology may themselves be negated if a SAL Treaty is signed in which these *low-unit-value* weapons are counted as equal to ICBMs, heavy bombers and SLBMs on a one-for-one basis. While a cruise missile *force* may well be much more effective than an equal-cost ICBM, SLBM, or bomber force, the single cruise missile is a much less powerful weapon than the single heavy bomber or the single ICBM. For this reason the imposition of ceilings in SAL in which cruise missiles are treated as one-weapon units, may have the effect of undermining all the potential usefulness of this advanced technology. At the time of writing it is uncertain what the SAL negotiations may yet bring; it is not known if cruise missiles will in fact be treated as one-weapon units, nor if cruise-missile *carriers* will be constrained under the multiple-warhead weapon sublimit.

A less economical and less innovative potential new weapon is the mobile ICBM. The United States tried and failed to include the prohibition of land-mobile ICBMs in the text of the SAL-1 Interim Agreement. Instead, on May 20, 1972, the United States delegation issued a "unilateral statement" according to which the deployment of land-mobile ICBM launchers was deemed to be "inconsistent" with the objectives of the agreement. Since the Soviet Union did not concur in the statement it has of course no *legal* significance; since the United States is not willing to enforce the statement in its SAL diplomacy, it has no practical significance either.

Mobile ICBMs are much less vulnerable to high-accuracy ballistic-missile attack than fixed-site ICBMs (even if housed in protected silos) because present ballistic missiles must be pretargeted against a fixed point on the earth's surface prior to launch. In theory, a satellite "real-time" surveillance and acquisition system coupled to a remote (or homing) guidance system could enable an attacker to tract and destroy moving ICBMs also. But no such systems are now in operation and none are being procured for deployment; they may never become worthwhile since satellite platforms are themselves inherently vulnerable to attack.

Many configurations for mobile ICBMs have been suggested including ballistic missiles launched from large transport aircraft held on ground or airborne alert (as recently tested in the United States); open-ended land-mobile deployments on road or rail (which would present obvious security problems); and, closed-circuit land-mobile deployments, where ICBMs would rotate from shelter to shelter inside launcher containers (not all of which need contain missiles) mounted on road or rail platforms. In the United States there are still some large military reservations which could be used for the purpose, and of course both the Soviet Union and China would have no difficulty in finding vast and secure deployment areas. (Any closed-circuit system would have to occupy a large land area since the mobile ICBMs would otherwise be vulnerable to area-wide barrage attacks.)

In designing a mobile ICBM system, the degree of shelter hardness (trench-like horizontal silos have been proposed) can be traded off against the extent of the deployment area, though the hardness of present silos would set an upper *cost* limit, while the need to preclude barrage attacks would set a vulnerability lower limit. Anything that moves is likely to have greater maintenance problems than a static system; there would also be some further complications in the design of guidance and control systems. These two factors are likely to compound the inherently high cost of mobile ICBM deployments. For this reason it is not certain that land-mobile deployments will remain an attractive solution to the ICBM vulnerability crisis which is expected to emerge once *present* Soviet ICBMs are deployed in large numbers with the sort of accuracies that may reasonably be projected for the mid-1980s.

The American objection to mobile ICBM deployments, expressed in the post-SAL-1 unilateral statement, was *not* motivated by one-side strategic considerations (the United States is technologically better placed to deploy mobile ICBMs than the Soviet Union) but rather by generic arms control considerations: mobile ICBMs cannot be counted. Nevertheless the Soviet Union has now proceeded with the *development* of the SS-16 (see Table 1) which seems to be designed for a mobile deployment. Further, the Soviet Union has actually deployed the SS-20 in a mobile configuration. With this, the mobile ICBM threshold may already have been passed, since with a light warhead the SS-20 does have an intercontinental potential.

Of late there has been a tendency to view strategic innovation as an inherently negative phenomenon. Indeed the current thrust of the arms control interest is manifest precisely in opposition to all forms of innovation (through testing bans). But the example of the cruise missile shows that innovation may offer the possibility of achieving the *substantive objectives* of arms control: if cruise missiles replace some of the present weapons in the strategic inventory, the result would be to reduce the overall cost and increase the overall stability of strategic forces. It is true of course that the development of cruise missiles also militates against the exterior forms and formal procedures of arms control *negotiations*, since cruise missiles are too small (and cheap) and too adaptable to be counted reliably in a SAL context. We have therefore a choice between deployment, which would achieve both strategic and arms control objectives but which would preclude treaty-making; and non-deployment, which would allow treaty-making but preclude the possibility of reducing the costs and increasing the stability of the strategic balance.

III. MEASURING STRATEGIC POWER

For all their sinister novelty, strategic-nuclear forces share the multiplicity of attributes which are common to all forms of military power. For this reason purported assessments of the strategic balance based on comparisons of any one attribute of the *materiél* are at best inadequate and at worse deliberately misleading. For example, comparisons of the *number* of delivery vehicles—as routinely presented in official statements (and shown in Table 8)—do *not* meaningfully define the balance of strategic power. They merely provide information on *one* aspect of vulnerability and operational flexibility: the greater the number of separate vehicles the greater—other things being equal—the overall survivability of the force and its flexibility in use.

But of course other things are not equal. One obvious difference is the greater progress the United States has made in payload subdivision through the deployment of multiple warhead systems, where each warhead can be aimed at a separate target. Indeed warhead (and bomber weapon) counts are sometimes offered as *overall* indicators of strategic power on the grounds that "it is warheads that kill people," to quote the simple-minded and misleading slogan which has now been officially sanctioned by presidential use. If a simplistic slogan such as this is found to have an irresistible attraction, then let it be "it is *surviving, reliable* warheads that kill people." In the continuing debate over strategic policy, those who oppose greater armaments and those who seek to defend the SAL accords have all tried to promote the warhead comparison—as opposed to the launcher comparison in which the Soviet Union is superior. But of course the warhead comparison, as shown in Table 9, cannot measure strategic capability since the warheads thus compared are not at all equal in yield or accuracy.

TABLE 8

Strategic Delivery Vehicle Comparison–Mid-1976 Estimate

	United States	Soviet Union
ICBMs, old	54	124
ICBMs, light	1,000	1,091*
ICBMs, heavy	0	285
SLBMs on nuclear submarines	656	825
SLBMs on diesel-electric boats	0	60**
Operational Manned Bombers	382	180
Manned Bomber Inventories	605	355
Total Vehicles	2,315	2,740
Total SAL-accountable vehicles	2,131	2,505+

*Includes SS-19s.
**Some diesel boats carry accountable modern SLBMs.

The one attribute of strategic power that the warhead comparison does validly measure is the operational flexibility of the force: other things being equal, the force with the greater number of independently targetable warheads will more effectively cover the enemy target system. Another purported measure of strategic power is the "total megatonnage" comparison, in which the combined energy yields of all warheads and bombs in the strategic inventories is simply summed. Often used for polemical purposes by those who seek to magnify Soviet strategic capability (in arguing for greater American armaments), the total megatonnage comparison is even more misleading than the warhead count. According to this method of measurement a single SS-9 warhead is estimated as the equivalent of 500 or more Poseidon warheads, and for that matter an

TABLE 9

Separate Warhead and Bomb Comparison–Mid-1976 Estimate

	United States	Soviet Union
ICBM warheads (MIRV and single)	2,154	1,980
SLBM warheads	5,120*	885
Air-ground missiles and bombs	2,550	360-460
Total Force Loadings (maxima)	9,824**	3,225-3,325
Total Force Loadings (official estimate)	8,900	3,500

NOTE: Deployed warheads, attack air-ground missiles and bombs on operational delivery vehicles in strategic deployment.

*Includes 640 SLBM warheads which will not be actually operational until 1978.
**Includes high estimate of bomber weapon loadings.

American Mk.28 bomb counts as the equivalent of several smaller Soviet ICBMs. It is plain that this sort of comparison is wholly misleading:

(1) The single SS-9 launcher is altogether more vulnerable than the 3.125 submarines which contain the 50 Poseidon SLBMs and their 500 warheads;

(2) The reliability of the Poseidon warheads is collectively very much greater;

(3) The operational flexibility of the single SS-9 warhead is a fraction of that a megaton-equivalent Poseidon force, which could attack as many as 500 separate targets.

It will be appreciated that the ability to attack selectively is an important attribute in deterrence: the *political* credibility of a limited strategic threat is conditioned by the degree to which it is truly limited. For example, a specific strategic threat against, say, Soviet submarine bases in the Kola peninsula is much more likely to achieve its object (that is, the deterrence of some equally specific Soviet act, such as an invasion of Northern Norway) if the collateral damage to Soviet civilian life is minimized. Otherwise it would not be difficult for the Soviet Union to negate the threat by itself threatening to attack an American city. The *political* versatility of a strategic force is dependent on its ability to inflict not maximum but rather *controlled* damage, a function of the availability of a wide range of different weapon yields (including low-yield weapons) and even more, of accuracy, as shown in the comparisons below.

Aside from these broader considerations–which between them do cover the real attributes of strategic power–the total megatonnage comparison is also misleading for a much simpler physical reason: since almost all targets are on a horizontal plane on the earth's surface while the destructive effect of detonations is diffused spherically in all directions, the real destructive effect of nuclear detonations does not increase proportionately with their energy yield. Targets are necessarily "over-destroyed" at the core of the explosion, so that much of the energy yield of explosive weapons is wasted. A cluster of many small explosions is therefore much more efficient against a given target area than a single detonation which has the same total yield.

A simple formula (2/3 power of the energy yield) can be used to convert the gross nominal yield of nuclear weapons as stated in megatons or kilotons into one-megaton equivalents (o.m.e.), the latter being a rough measure of real destructive power against "soft" target areas, such as cities. For example, a 25 megaton weapon (roughly the yield of the very largest Russian ICBM warheads) has an o.m.e. of 8.55. In other words, it can destroy just over eight and half times as much as a one megaton weapon, which nominally has only 4 percent of its yield. Similarly, Poseidon warheads have a nominal yield of only 40 kilotons (equivalent to 40,000 tons of T.N.T.) but their o.m.e. yield is 117 kilotons. This means that 10 Poseidon warheads could in fact destroy a greater target area than a single one-megaton weapon, which has more than twice their combined nominal yield.

A greater degree of yield subdivision into separate warheads is normally more efficient, and the availability of low-yield warheads enhances versatility, but large-yield warheads are useful also. To attack point targets which are heavily protected, such as command and communications centers as well as ICBM silos, large-yield warheads are not merely better than low-yield ones but they are also superior to *more accurate* low-yield warheads whose theoretical lethality is just as great. Late-model upgraded American ICBM silos are estimated to have a resistance to peak overpressure of up to 1,000 pounds per square inch (psi). In theory, the weapons of varying yield and accuracy, as shown in Table 10, all have a 95 percent single-shot probability of destroying a 1,000 psi silo.

In theory, therefore, a 170 Kt warhead such as now fitted in Minuteman-3 warhead systems will be as lethal as a 20-megaton warhead as deployed on Soviet SS-9s, when the accuracy of the two missiles is 0.065 and 0.33 of a nautical mile respectively. However, the latter accuracy *is* fully attainable with present guidance techniques, while the former is not. Further, while yields can be determined quite accurately by the deploying party, accuracies can only be estimated: the data can only be obtained by observing missile tests, and reliable estimates would require many full-range tests from normal operational launchers *in the direction of intended attack.* Present accuracy estimates must instead be based on few tests from special test launchers (operated by special maintenance crews) and the missile obviously cannot be sent in the direction of the intended target. (Both the Soviet Union and the United States launch test missiles into the Pacific and not at each other.) Moreover, in operational conditions there is every

TABLE 10
Alternative Weapons with an Equal Single-Shot Kill Probability

Yield in Megatons:	0.04	0.17	1.0	5.0	20.0
Required Accuracy (C.E.P.) in nautical miles	*	0.065	0.12	0.25	0.33

*The required accuracy is less than 0.05 of a nautical mile.
C.E.P. = Circular Error Probable = radius within which half the warheads are expected to fall.

reason to believe that accuracy will be more sharply degraded than warhead yield: in fact the latter should not be affected at all unless a catastrophic failure takes place.

The two arsenals can on the other hand be usefully compared in terms of the yield choices offered by missile systems. This will be an indicator of operational flexibility: low-yield warheads are recognized as particularly useful for selective strikes against soft targets while high-yield warheads are generally more useful for "counterforce" attacks against targets which are compact and heavily protected, such as ICBM silos (see Table 11).

It is not therefore accurate to say that the United States has the possibility of attacking many more targets than the Soviet Union; rather the United States can attack many more *soft* targets, while the Soviet Union is better placed to attack heavily protected targets, including ICBM silos. This observation must be modified, however, by the known fact that American strategic missiles are more accurate than those of the Soviet Union. It is very hard to estimate accuracies even when full access to all the data is available. As it is, the figures given in Table 12 are no more than informed guesses. It is wholly inadmissible to treat these as hard facts for the purposes of strategic calculations which seek to achieve more than suggestive results.

Estimates of expected accuracy can be combined with energy yield figures by using the well-known "counterforce" equation (two-thirds power of yield divided by the square of the CEP), and the resulting "lethality" can then be stated in a single figure. This can be compared in turn with the "lethality" required to destroy any type of target so that it would appear that an entire strategic scenario can be worked out by merely comparing two sets of figures: force-wide lethality viz. the target-wide lethality

TABLE 11

Energy Yields and Operational Flexibility—Mid-1976 Estimates

Number of Separate ICBM/SLBM Warheads by Yield Categories

	Under 50 Kt.	50-200 Kt.	201-999 Kt.	1-4 Mt.	5-10 Mt.	Over 10 Mt.
United States						
ICBMs	0	1,650	0	450	54	0
SLBMs	4,960	–	160	0	0	0
Soviet Union						
ICBMs	0	0	580	991	124	285
SLBMs	0	0	0	885	0	0

Kt. = kiloton = 1000 tons of TNT equivalent.
Mt. = megaton = million tons of TNT equivalent.

TABLE 12
Estimates of Expected Missile Accuracies (Current Forces)
In nautical miles or fractions, Circular Error Probable

United States			
Minuteman-2 = 0.3	SS-7 = 2.0	SS-19 = 0.2-0.3	SS-N-6 = 1.0
Minuteman-3 = 0.2	SS-8 = 2.0	SS-9 = 0.7	SS-N-8 = 0.8
Titan-2 = 0.5	SS-11 = 0.8	SS-18 = 0.2-0.3	
Polaris A3 = 0.7	SS-13 = 0.8	SS-N-4 = 1.0	
Poseidon C3 = 0.3	SS-17 = 0.2-0.3	SS-N-5 = 1.0	

requirement. But there seems little point in the exercise: strategic arsenals consist of thousands of complex weapons manned by tens of thousands of fallible men. Hence allowance must be made for weapons in maintenance on a routine basis (though forces can of course be "peaked"), and further and less certain allowances must also be made for technical failures. Lethality calculations ignore this. More important, a counterforce strike is not an automatic exchange of "lethality" for protection, but rather a complex military operation whose outcome is critically affected by the possibility of timely replacement for weapons which are observed to fail (reprogramming), the timing of the attack, the allocation of forces to targets, and so on. By contrast the theoretical calculation of lethality is apt to lead to results which are operationally impossible, such as the simultaneous attack on single targets by several warheads (they would destroy one another), and the destruction of many targets by a single very lethal warhead.

The impact of some of these operational realities is illustrated in Table 13.

It will be noted that after modest allowance has been made for ICBMs in maintenance and for technical failures, the present American ICBM force could not be usefully employed in an offensive against the Soviet ICBM force, *even under theoretically optimal* operational assumptions: even if Soviet ICBM silos are only protected against peak overpressures of 300 psi, no more than 871 out of 1,500 would be destroyed. In the wake of the attack, none of the American ICBM force would remain (except for ICBMs in maintenance) while 629 ICBMs would still remain in the Soviet arsenal. If Soviet ICBM silos are better protected (as they should be, since they are on average more modern than U.S. silos) the post-attack outcome is again zero for the United States versus 823 for the Soviet Union. And this of course is the essential mechanism of strategic stability: a balance is stable when it does not pay to launch a would-be disarming counterforce attack.

TABLE 13
American ICBM Forces: Theoretical Anti-Silo Capabilities
(1976 Assumptions)

	Titan-2	Minuteman-2	Minuteman-3
Deployed ICBMs	54	450	550
ICBMs operationally available (0.9)	49	405	495
ICBMs reliable through boost phase (0.85) (0.85)	42	344	421
Warheads on reliable ICBMs	42	344	1,263
Warheads reliable through MIRV separation (0.9)	42 (n/a)	344 (n/a)	1,137
Warheads reliable on target (0.95)	40	327	1,080
Kill probability against:			
300 psi Soviet ICBM silos	84%	71%	56%
500 psi Soviet ICBM silos	74%	56%	43%
Theoretical Maxima			
300 psi silos destroyed	34	232	605
500 psi silos destroyed	30	183	464

Even this result grossly overstates the counterforce capability of the American ICBM force, for it was implicitly assumed that *all* reliable U.S. warheads would somehow be unfailingly targeted on undestroyed Soviet silos. In reality, an offensive could not be perfect, as in the synthetic offensive outlined in Table 13. Perfection would require

(1) the successive allocation of available warheads to replace failed warheads through "reprogramming"; and

(2) the avoidance of interference ("fratricide") between two or more warheads impacting in the same target area.

Reprogramming presents no great difficulty when it comes to compensating for missiles which are observed failing at launch or in the boost phase (the first few seconds of flight); but it is virtually impossible with present equipment to allocate additional warheads on a one-for-one basis to replace warheads which fail *after* the boost phase. As for the problem of fratricide, it will be intuitively obvious that the immediate effects of a nuclear detonation and the column of debris it throws upwards will destroy or at least deflect incoming warheads that follow closely. And if successive waves of attack do *not* follow closely, then there is every likelihood that the victim will fire his own surviving ICBMs in between the waves of attack (or rather continuously, but only those fired *between* the attack waves would survive).

A full scale analysis which incorporates these operational realities cannot be usefully presented here. But a simple illustration of a quasi-realistic 1985 counterforce attack may suggest the rough limits of the capabilities of the two sides (see Table 14 below). It may be seen therefore that even under near-optimal operational conditions, with two attack waves and early-phase failure reprogramming, an American counterforce attack is still not a paying proposition: all American ICBMs are expended, while 396 Soviet ICBMs survive. The counterforce potential of American SLBM warheads would in theory suffice to destroy most of these remaining war-

TABLE 14
American ICBMs Expended Against Soviet ICBM Silos: 1985 Assumptions*

	Minuteman-2	Minuteman-3
1st Wave		
Deployed ICBMs	440	560
ICBMs available (0.9)	396	504
Warheads available	396	1,512
Of which assigned to targets in 1st wave	–	1,470**
Warheads reliable through boost phase	–	1,251
Warheads reliable through impact	–	1,070
Kill probability against 1000 psi silos	(65%)	86%
Soviet silos destroyed: 1st wave	–	920
Reprogrammed ICBMs (one-for-one for launch and boost-phase failures)	177	14***
Warheads available	177	42
Of which reliable through boost phase	150	36
Warheads reliable through impact	143	31
Soviet silos destroyed: 1st wave reprogramming	93	27
2nd Wave		
Residual available ICBMs	219	0
Residual available warheads	219	–
Of which reliable through boost phase	186	–
Of which reliable through impact	177	–
Soviet silos destroyed: *net* increment****	34	–
Total Soviet silos destroyed by type	127	947
Force-wide Total	1,074	

*1985 Assumptions: Minuteman-2 = one 1-Mt. warhead 0.2 n. mile CEP;
**Minuteman-3 = three 350 Kt. warheads 0.11 n. mile CEP;
***limited by the number of targets;
****177 terminally reliable warheads destroy 115 silos of which 75% already attacked by MM-3 warheads (Net kp = 0.65 x 0.14); of which 16% unattacked owing to terminal failures, and 9% attacked by 1st wave MM-2 warheads (Net kp = 0.65 x 0.35).

heads, but the operational problems of launching a coordinated ICBM/SLBM attack against such time-sensitive targets could not be overcome unless all manner of equipment is simply assumed into existence.

But of course this estimate does not mean that strategic stability will in fact be preserved, for that depends equally on the counterforce capabilities of the Soviet ICBM force. Under identical assumptions, these counterforce capabilities can be estimated as follows:

TABLE 15
Soviet ICBM Attack on American ICBM Silos: 1985 Assumptions*

	SS-17	SS-19	SS-18 Mod. 2**
1st Wave			
Deployed ICBMs	516	516	288
Warheads available	1,858	2,786	2,074
Of which assigned			1,000
Of which reliable (I) early phase			850
Of which reliable (II) late phase			727
Kill probability against 1000 psi silos	(50%)		65%
American silos destroyed: 1st wave			473
Reprogrammed warheads			150***
Of which reliable (I)			128
Of which reliable (II)			109
American silos destroyed: 1st wave reprogramming			71
2nd Wave			
Residual available ICBMs	516	516	115
Warheads available	1,858	2,786	920
Warheads assigned	80		920
Of which reliable (I)	68		782
Of which reliable (II)	58		669
American silos destroyed: 2nd wave (Net)****	11		171****
Reprogrammed warheads	150		
Of which reliable (I & II)	109		
American silos destroyed	20		
Total American silos destroyed by type	31		715
Force-wide Total		746	

*SS-18 MIRV model: eight one-megaton warheads with a 0.2 n. mile CEP. SS-17: four 0.5 Mt. warheads with a 0.2 n. mile CEP.

**Or follow-on type.

***Actually 152, but 2 wasted.

****The number of silos destroyed which were not destroyed earlier: 669 reliable SS-18 warheads destroy 435 silos including 372 silos already attacked of which 242 were already destroyed, and 63 silos unattacked owing to technical failures in terminal phases.

It may appear that the results of the synthetic Soviet counterforce attack outlined in Table 15 are similar to the results of the American attack, since in both cases the enemy force is not fully destroyed. But there is a fundamental difference: while the American ICBM inventory was fully expended in launching the attack, only part of the Soviet ICBM force was used. As a result, the Soviet Union *would be better placed after an attack than before it.* Thus, while the United States would have no incentive to launch a counterforce offensive, the Soviet Union would (see Table 16). In other words if the United States strikes first, its position deteriorates with the margin of Soviet numerical superiority increasing almost threefold. On the other hand, if the *Soviet Union* strikes first, its numerical superiority increases by 300 percent. Thus the fundamental mechanism of a strategic stability is compromised.

What makes these results so disturbing is the nature of the assumptions made in each case. While the American force was credited with maximal performance (given present technologies), Soviet accuracies were assumed at 1976 U.S. levels, thus discounting the ability of the Soviet Union to do better even after a full decade of development work on its *new* ICBM systems. (An entire family of follow-on ICBMs is now in the R&D phase in the Soviet Union.) The yield-to-weight ratios assumed were also on the low side (the SS-18 warheads may easily exceed one megaton, and the SS-17 warheads could be as much as one megaton). But higher yields would add nothing to the counterforce capabilities of the Soviet ICBM force: it already contains more than enough extra-large multi-megaton warheads to cover every conceivable heavily protected target other than ICBMs, while the latter could not be usefully attacked by large yield weapons owing to

TABLE 16
Strategic Stability Under 1985 Assumptions

	U.S. ICBM Force	Soviet ICBM Force
1. United States Counterforce Attack:		
Pre-Attack Inventory	1,000	1,470
U.S. Force as % of Soviet Force	68%	
Post Attack Inventory	90	396
U.S. Force as % of Soviet Force	23%	
2. Soviet Counterforce Attack:		
Pre-Attack Inventory	1,000	1,470
Soviet Force as % of U.S. Force		147%
Post-Attack Inventory	254	1,153
Soviet Force as % of U.S. Force		454%

the reciprocal interference of the detonations. Subject to this and other operational problems (which are *not* wholly incapable of solution), it is apparent that with fair accuracies (attainable *without* terminal guidance) the 1985 Soviet ICBM force would be able to destroy all but a fraction of the Minuteman silos. And it is this prudent estimate which must guide American defense planners, given the long lead times of deployment and the high political and financial costs of remedial action in crash program conditions.

While the relentless advance of some technologies have prejudiced the survivability of fixed-site, land-based ICBMs, the very great investment made since the 1950s in anti-submarine technologies has not resulted in truly major innovation. Hence the submarine basing mode still offers a reliable guarantee of survivability if all required safeguards are observed. Well-designed and well-handled nuclear-powered submarines on quiet patrol cannot be detected reliably by any known ASW technique. This state of affairs may change if radically new technologies are introduced, but so far the peculiar properties of sea water have defeated all attempts to develop reliable means of accurate long-range location. Current ASW forces are therefore useful primarily against "attack" submarines which must reveal themselves in striking at their targets.

It has been suggested that missile submarines, being few in number, could become vulnerable to "trailing tactics," where attack boats would follow SLBM boats as they leave their refitting bases to pursue them thereafter throughout their cruise. In this manner, the SLBM force at sea could be kept within torpedo range continuously, with the attack boats ready to fire on command. In practice there would be severe difficulties: first, the exits of SLBM boats can be masked by contrived noise; second, if they move slowly SLBM boats will generate noise levels which remain below the loudness level of the surrounding sea. This would ordinarily defeat sonar, virtually the only means of detection available to an attack boat operating in such circumstances. Finally, trailing tactics would require the permanent deployment of two or three attack boats for each SLBM in the adversary force. Neither the United States nor the Soviet Union have enough boats to use the tactic against each other.

Given the current state of ASW techniques the net SLBM strike capabilities are largely defined by logistic factors. In the Russian SLBM fleet, however, there are still some 20 diesel-electric boats which are noisy and which must come relatively close to American shores to launch their missiles. These boats must be judged as highly vulnerable. The H-class and Y-class submarines in the Soviet fleet would not expose themselves through the use of air-breathing gear since they are nuclear-powered, but they too are reputed to be significantly more noisy than American Polaris submarines and the operations of H-class boats would further be limited by the range of the SS-N-5 missile. The D-class may or may not be noisy, but their long-range SS-N-8 missile should ensure virtual immunity from detection in many millions of square miles of sea. It may be assumed that all submarines at sea would eventually reach their firing stations to launch their attack, while submarines in port would all be lost to attack. Hence only the former can be taken into account in computing minimal-assured strike-back capabilities (see Table 17). The 344 o.m.e. of the survivable/reliable American SLBM force could in theory destroy roughly one-quarter of the population of the Soviet Union (given nonimplementation of Soviet civil defense plans) and rather more than two-thirds of Soviet industrial capacity. Given American population densities and the abandonment of civil defense, the destructive capacity of the Soviet SLBM force is greater still. In practice, it is "industry" and not population that should serve as a benchmark for retaliatory attacks since populations are capable of being

TABLE 17

American and Soviet SLBM Strike-Back Capabilities
1976 static assumptions (compare with Table 5)

	Inventory	Max. at Sea	ASW Survivable	Reliable	
United States					
Polaris A-3s	160	80	80	64	
Poseidon C-3s	496	240	240	240	
Total Warheads					1,984
Total Equivalent Megatons					344
Soviet Union					
On G-class: SS-N-4s	27	12	0	–	
SS-N-5s	33	15	9	7	
On H-class: SS-N-5s	21	12	6	5	
On Y-class: SS-N-6s	544	272	224	179	
On D-I class: SS-N-8s	132	60	60	48	
On D-II class: SS-N-8s	128	64	64	51	
Total Warheads					290
Total Equivalent Megatons					347

Notes:

Reliable = 0.8 overall degradation factor applied across the board. (In reality MIRV systems and older systems should be less, and non-MIRV modern systems should be more warhead reliable.)

Equivalent megatons: Poseidons computed at 0.117 o.m.e./per warhead @ 10 warheads per missile; Polaris A-3s MIRV system computed at 0.99 o.m.e.; all Soviet SLBM warheads except SS-N-8s assessed at one metagon (two is more likely) except for SS-N-8 assessed at two megatons, = 1.58 o.m.e.

evacuated. As the least vulnerable weapons in the nuclear arsenal, SLBM forces are likely to be retained till the end, even in the wake of full scale ICBM attacks (especially if these are counterforce strikes which should *not* kill large proportions of the population). Hence there might be ample time for pre-attack evacuation in the Soviet Union, whose civil defense precautions are costly, very elaborate, and certainly not ineffectual. While the post-attack prospects of Soviet society would still be catastrophic by any standards, the immediate casualty toll that the American SLBM force could inflict on evacuated Soviet cities might not exceed 10 percent or so of the Soviet population.

Land-based ICBMs are vulnerable prior to launch but not in the approach to the target; SLBMs are vulnerable mostly to attacks on their bases but they are not significantly vulnerable either at launch *or* penetration. Bombers are by contrast vulnerable both at their launch and in penetration to their targets. And yet they are still potentially the least vulnerable of all strategic weapons over time, since only bombers are flexible enough to adjust to changes in the threat. Even "supersonic" bombers would fly intercontinental missions at subsonic speeds. Given the several hours that home-based American or Russian bombers would need to reach each other's homelands, it is generally assumed that bombers would have no useful counterforce role. (But Soviet bombers could certainly be used against Chinese MRBMs, which reportedly cannot be launched without lengthy pre-fire preparations.) It follows that the *strategic* capabilities of manned bombers can be evaluated exclusively in terms of limited attacks and retaliatory strike-back attacks. For the latter, the important variables are:

(1) the ability of the bombers to survive would-be disarming strikes upon their bases, and

(2) the ability of the bombers to penetrate air defenses.

Bombers on the ground are neither protected as ICBMs are, nor concealed as the SLBMs. Instead their survival depends on their ability to take off rapidly to leave the base zone altogether after receiving early warning of impending attack. Further, in order to penetrate hostile airspace at low altitude, bombers require midair refueling, and so the overall effective survivability of the bomber force against missile attack depends not only on the efficiency of warning systems and bomber take-off readiness, but also on tanker survivability.

If the need arose, the survivability of both bombers and tankers could be increased at short notice by placing a proportion of the force on continuous airborne alert; this has been done in the past and it could be done in the future though at great cost. Until recently, 40 percent of the U.S. bomber force was kept on a 10-15 minute ground alert, this time-lag being predicated on a slightly longer early-warning interval. Under the practices introduced in 1975 only about 30 percent of the force is kept on ground alert. Again, this proportion could be increased to more than 70 percent for a short-term surge; but unless USAF manning levels were increased, this high rate of alert could not be sustained for long. The effectiveness of ground alert as a survival tactic depends on the length of the pre-attack warning. Both the Soviet Union and the United States could in theory strike at each other's bomber bases with depressed-trajectory SLBMs which would cut warning times. Travelling at higher speeds over shorter ranges, such specially configured SLBMs could reach bomber bases even deep inland in the United States in eight minutes or so. They would therefore compromise the value of present ground alert practices on the present base system (which is not optimally located). If such SLBMs were indeed deployed, the survival of the bomber force would require the adoption of crisis alert procedures or even an airborne alert. American bombers are deployed as an *independent* strategic force, on the presumption that the strategic value of the bombers depends on their ability to penetrate enemy air defenses *without* having the way opened for them by prior ICBM or SLBM attacks. Since current Soviet air defenses are still optimized for the interception of high-speed/high-altitude bombers (such as the abortive B-70 of the early 1960s would have been), it is generally thought that the old and rather slow B-52s should not suffer heavy losses to defending SAMs and fighters. Soviet bombers are much less well-equipped to defeat interception but on the other hand, the effectiveness of American air defenses is severely limited by the very small number of manned interceptors and the absence of "point" missile defenses. It is generally expected that the Soviet Union will persevere in the attempt to develop airborne radar control for anti-bomber defense. This would be a counter to low-level penetration (a tactic itself intended to counter ground-based radar). Against a defense equipped with effective look-down radars and snap-down missiles, present flight tactics would not suffice to ensure acceptable rates of penetration. Instead, an aircraft like the B.1 could fly much lower still in a "contour-flying" (terrain-following) mode at high subsonic speeds, in order to seek concealment in ground clutter. Moreover, using their high-speed/high-altitude flight capability some B.1s could be assigned to the attack of airborne air defenses with long-range air-to-air missiles. (Hence the deployment of the B.1 would force the Soviet Union to deploy both high- and low-altitude SAMs, as well as the airborne air defenses.) A range of estimates for bomber strike-back capabilities is presented in Table 18.

TABLE 18

American and Soviet Bombers Strike and Strike-Back Capabilities

	United States			Soviet Union		
	B 52 D	B 52 G/H	FB-111	Tu-95	Mya-4	"Backfire"
Aircraft in operational units*	75	241	66	105	35	40
Number normally on ground alert	23	72	20	(32)**	(11)	(12)
Crisis alert deployment	53	169	46	74	25	28
Flight reliability	0.9	0.9	0.9	0.7	0.7	0.9
Proportion surviving air defenses						
High estimate	0.7	0.8	0.9	0.7	0.7	0.8
Low estimate	0.5	0.6	0.8	0.6	0.6	0.6
Weapon loadings, o.m.e. units***	4.0-12.0	6.0-14.0	0.7	2.0	2.0	2.0
Total deliverable weapons						
Crisis alert/low combat loss	134-400	730-1,704	26	73	25	40
Normal alert/high combat loss	41.4-124	233-544	10	27	9	13
Force-wide totals, Maximum		2,130			138	
Minimum		284			39	

*From Table 6. **Not normally kept on ground alert. ***Based on data of Table 7.

Of necessity the defense attrition rates assumed here are highly speculative. For all the limitations of the data, it is legitimate to interpret Table 18 to mean:

(1) the American bomber force is much more capable than the Soviet—and would remain so even if all U.S. air defenses were scrapped; and

(2) bomber forces in general are much more flexible than other forms of strategic power.

It is apparent that a heightened alert status is more than a symbolic diplomatic signal, for effective retaliatory capabilities may thereby be increased by a factor of two or more.

What all these evaluations of strategic power have in common is the sharp divergence between the nominal capabilities of strategic inventories and their *expected* capabilities in plausible operational conditions. It should be obvious that in *real-life* operational conditions in nuclear environments the divergence between theoretical and actual operational capabilities might be very much greater still. For this reason, simplistic comparisons of strategic *inventories* are grossly misleading. In the so-called "overkill" scenario for example, the overall equivalent-megatonnage of American *inventory* forces is compared to the level deemed necessary to destroy the major cities of the Soviet Union; it is then concluded that the United States has enough "strategic power" to destroy the Soviet Union many times over. As soon as the contingency of enemy attack on the strategic force is taken into account, as soon as the technical degradation of the force is allowed for, we find that the "overkill" rapidly diminishes. In fact if Soviet civil defense planning is taken at face value, perfectly realistic operational assessments show that far from being "destroyed many times over," the Soviet Union might not lose more than 10 percent or so of its population in a full-scale American strategic attack. This does not of course mean that the overall destructive capability of American strategic forces is inadequate. It merely reflects the inherent difficulty of destroying dispersed populations, and it does suggest that American civil defense planning may deserve greater attention than it now receives.

Enough has been done to allow the reader to make his own estimates of strategic power. Above all, it should be clear that no comparison of the *materiél inputs* can be at all adequate to define the strategic balance. It is rather the *output* of operational capabilities which must be compared, in a setting as realistic as the available information will support.

Glossary

ABM TREATY	Anti-Ballistic Missiles Treaty
ALCM	Air-launched cruise missile
ASM	Air-to-surface missile
ASW	Anti-submarine warfare
AWACS	Airborne warning and control system
CEP	Circular error probable
ECM	Electronic countermeasures
FBS	Forward-based systems
FOBS	Fractional orbital ballistic system
ICBM	Intercontinental ballistic missiles system
IOC	Initial operational capability
IRBM	Intermediate range ballistic missile
KM	Kilometers
KT	Kiloton fission warheads
MBRM	Medium range ballistic missile
MIRV	Multiple Independent Reentry Vehicles
MK 12A WARHEADS	New Minuteman warhead system with larger yield warheads
MRV	Multiple reentry vehicle
N. MILE	Nautical mile
NORAD	North American (U.S.-Canadian) Air Defense Command
O.M.E.	One-megaton equivalents
RV	Reentry vehicle
SAF	Soviet Air Force
SAL-1	Strategic arms limitation talks
SALT	Strategic arms limitation talks
SAMS	Surface-to-air missiles system
SLBM	Submarine-launched ballistic missile
SRAM	Short-range attack missiles
TERCOM	Terrain Reference System
THROW-WEIGHT	Missile payload deliverable over any given range
WS. 120	An abortive U.S. Air Force Project for a large capacity ICBM

Bibliographic Note

Any serious attempt to engage in the strategic discourse must begin with an understanding of weapon effects. For this, Samuel Glasstone (1964) *The Effects of Nuclear Weapons*, U.S. Gov. Print. Office, rev. ed. is basic. A generic discussion of weapon types is found in my *Dictionary of Modern War* (1972), New York: Harper & Row, naturally not fully up to date. Among the simple manual aids to strategic calculations, perhaps the best is the *Missile Effectiveness Calculator*, produced by the Heavy Military Electronics Division of General Electric. (The *Rand* pinwheel is simpler and handier, though perhaps less enlightening.) An up to date general introduction to the subject in brief is found in Normal Polmar (1975) *Strategic Weapons: An Introduction*, New York: Crane, Russak, primarily useful for the historical background. A good outline to the policy issues is found in Alton H. Quanbeck and Barry M. Blechman (1973) *Strategic Forces: Issues for the Mid-Seventies*, Washington, D.C.: Brookings; the one major issue not there mentioned, the power-perception nexus, is covered well in the "posture statements" of former Secretary James R. Schlesinger, especially the (1975) *Annual Report of the Department of Defense*, also found as *Statement of James R. Schlesinger before the Armed Services Committees of Senate and House for Fiscal Year 1976*. An important background document is the 1972 volume of hearings on the SAL-1 accords (Military Implications of the Treaty on the Limitations of Anti-Ballistic Missiles Systems. *Hearings* before the Committee on Armed Services, United States Senate, 92nd Congress, 2nd Session, 1972). By far the most important contribution to the strategic debate of the last several years is Albert Wohlstetter (1975) *Legends of the Strategic Arms Race*, Boston: U.S. Strategic Institute.